Robert W. Ayton

B.Ed. Mathematics Education

INTRODUCTION

This book covers all the Algebraic topics necessary for the BGCSE, GCE and CXC Mathematics curriculum. However, although this book was written for students in schools towards preparing them for these examinations; it was written with an in-depth informative take so that adults can use it to teach themselves.

The aim of this book is to teach the student to understand algebraic concepts and principles and applying them as they are learning. It focuses on honing and nurturing students' mathematical skills.

Teachers will find this book very helpful, using it as a supplement or with strategies in teaching this topic within the classroom.

This textbook, endeavours to assist students to overcome any weakness and fear they have for Algebra and to break that blockage and loose them from that bondage they had before encountering this textbook.

Exposure is given with algebraic terms and concepts by placing a glossary within the book for the student to be able to translate the information within the text.

Anyone who encounter this textbook will never be the same again. This is because they will find the book very stimulating as it will enroute unto the path of enlightenment.

PREFACE

Algebra: Plain and Simple was written specifically for students leading to BGCSE, GCE, CXC and all walks of life. It is viewed by many that mathematics is a challenging subject. Even adults up to this day speaks about the fear of Mathematics. The concepts of Algebra as a mathematics topic are even regarded as mind boggling. According to many; the x *or the* y; in general a bunch of letters that just doesn't make sense.

This book takes into account that nothing should be taken for granted. As such, this book delves into explaining the when, the where, the how and the why, so that key areas of knowledge, understanding and skills will be developed. Thus, resulting in providing a firm foundation to raise competence and boost confidence towards Algebra; the mother of mathematics.

The concise approach of this text is designed to make the study of Algebra interesting, enjoyable and sound. Notedly, many different types of examples with explained workings are laid out in this text; with a few alternate methods, so that students can acquire the necessary skills in order to solve the problems in the given activities and quizes.

It is hoped that this text will transcends all students into a deeper and richer understanding of Algebra and thus enable them to garner greater success in their examination.

Robert Ayton

ALGEBRA 1

CONTENTS

ALGEBRA 2

CONTENTS

TOPICS AND SUB - TOPICS SUMMARY

This book is divided into two sections:

- Algebra 1
- Algebra 2

Under each topic or sub-topics, the concepts are written; giving definitions, explanations and ideas of what to look for in order to comprehend the topics or sub-topics. Further explanations, and ideas are given in the examples with the aim of concretizing the concepts.

GLOSSARY:

- **Variable** – this is an alphabet or that which represents an unknown number or unknown value or unknown quantity. **Examples:** $\boldsymbol{x, y, \beta}$ *etc.*
- **Constant** – a specific number or value that never changes in expression. A fraction can also be a constant. **Example: 5, 3.14.** $\frac{3}{4}$ ***etc***
- **Coefficient** – is a number that is multiplied by a variable. This number works along with the variable by multiplication. It works alongside the variable but is always position before the variable. **Example: 2a, 5y**. The 2a and 5y come about by saying $2 \times a$ and $5 \times y$ which gives 2a and 5y respectively. Note, the 2 and 5 are coefficients of **a** and **y** respectively and that the coefficients are written first. So, if we were given p *x* 23 it will produce 23p. Again, the coefficient MUST be written first.
- **Expressions (algebraic expressions)** – any mathematical statement which consists of variables, numbers and an arithmetic operation between them.
  Example
  (1) $4m + 7$

  (2) $3c + 7d - \frac{2}{3}x$

  The mathematical operations includes **+, −, *x*, ÷, exponents and roots.**
- **Terms** – A term is either a single number or variable or product of several numbers and variables. Terms are separated by a plus (+) or minus (–) sign in an overall expression. **Example: Single term = 5y, 7 etc.** which are called *Monomial*; **Expressions with more than one term = 9p + 4**, which is called a *Binomial*. The two terms are 9p and 4. Noting that each term is separated by the **+ or −.** And 9p and 4 were separated by **+** sign in that example. Additionally, **5y + 7 – 4*x*y**, which is called *Trinomial* having three (3) terms. While having four (4) and more than four terms are known as *Polynomials*.
- **Equations** – an algebraic equation is where two expressions equal to each other. **Example:** if we are given two expressions such as binomial x + 5 and a monomial expression 0 and let them equal each other; such as x + 5 = 0; an equation is created. As such, equations are formulated when any terms (monomial, binomial, trinomial or polynomial) equal to anyone of each term.

Rules for multiplying and dividing positive and negative numbers:

Under Multiplication:

$$(-1) \times (-1) = --1 \times 1 = +1$$

$$(-1) \times (+1) = +-1 \times 1 = -1$$

$$(+1) \times (-1) = -+1 \times 1 = -1$$

$$(+1) \times (+1) = ++1 \times 1 = +1$$

Under Division:

$$(-1) \div (-1) = --1 \div 1 = +1$$

$$(-1) \div (+1) = -+1 \div 1 = -1$$

$$(+1) \div (-1) = +-1 \div 1 = -1$$

$$(+1) \div (+1) = ++1 \div 1 = +1$$

Note: $The - and + symbols\ \ that\ came\ after\ the\ \times and\ \div$ $signs\ were\ placed\ at\ the\ front\ along\ with\ the - or +$ $signs\ that\ were\ already\ at\ the\ front. This\ was\ done\ seeing\ that - and +$ $are\ not\ inverse\ of\ either\ the\ \times or\ \div signs.$

Note: $Whenever\ these\ signs:\ + \& - are\ right\ next\ to\ each\ other;\ such\ as:$ $+ - or - +, they\ produce\ a - sign.\ Whenever\ the\ signs\ are - -;\ next\ to$ $each\ other\ \ they\ produce\ a + sign. Whlie + +;\ next\ to\ each\ other\ remains\ +.$

Note: $n\ .n\ means\ n\ \times n.\ The\ dot\ means\ to\ multiply$

Translating Algebraic Expressions

Note: To translate any algebraic expressions, it requires the ability to take a word phrase and represent it using symbols as an algebraic expression and vice versa.

Example:
Instructions: Translate the following word phrase into algebraic expressions using the given symbols.

(1) Four times a number n
Working:
The key word **times** means to multiply. So, the question is asking to use the symbols 4 and n and multiply them. That is $4 \times n$ which gives 4n
Answer = 4n

(2) Nine times a number n plus a second number y
Working:
$9 \times n + y = 9n + y$
Answer = 9n + y

(3) Seven minus eight times p
Working:
$7 - 8 \times p = 7 - 8p$
Note: Order of Operation requires to multiply before we Add or Subtract
Answer = 7 – 8p

(4) The square of the sum of two numbers **m** and **n**
Note: You are required to **square** (raised to the power 2) both **m** and **n** after they are added (**summed**)
Working:
We sum **m** and **n** to give = **m + n**
Now squaring both m and n gives $(\boldsymbol{m} + \boldsymbol{n})^2$
Answer = $(\boldsymbol{m} + \boldsymbol{n})^2$

(5) A third the product of p and q
Note: product means to multiply
Working: $\frac{1}{3} \times p \times q$
Answer = $\frac{1}{3}pq$ or $\frac{1pq}{3}$

(6) A beverage set consists of a mug, a coffee pot and a serving plate. The mug costs p dollars. The serving plate costs $7 more than the mug. The coffee pot cost four times as much as the mug.

Expressing your answers in terms of p

(i) The cost of the serving plate

(ii) The cost of the coffee pot

Working:

Note: when asked to express your answer in terms of **p**; it means that the variable p MUST be a part of the expression given.

(i) As stated, the serving plate costs $7 more than mug. Therefore, if the mug cost p dollars and the serving plate costs $7 more. Then the word more indicates (+).

Answer = p + 7

(ii) This gives p × 4 = 4p

Answer = 4p

Activity 1a

Translate each of the following word phrases into algebraic expressions by using the given symbols.

(1) Five times a number y.
(2) A quarter the product of two numbers $c\ and\ d$.
(3) The square of the sum of two numbers x and y.
(4) Ten times a number x, plus a second number y.
(5) Eight times a number p, minus a second number q.
(6) The cube of the difference between two numbers $x\ and\ y$.
(7) Two times a number x, added to six times a second number y, divided by five times a third number z.
(8) The square of the product of $a\ and\ b$, minus the square root of c.
(9) The square root of the sum of $x\ and\ y$.

Activity 1b

Translate each of the following algebraic expressions into word phrases.

(1) $7y$ (2) $4m + 5$

(3) $9y - 8$ (4) $(x + 3y)^2$

(5) $\frac{2}{3}ab$ (6) $\frac{(a+b)}{5}$

Activity 1c

Answer each of the following questions below

1. Ashley bought skateboarding equipment. The safety pads cost a total of \$52. The skateboard cost \$$y$ more than the safety pads. The helmet cost \$17 less than the skateboard. Write an expression, in terms of y, for
   (a) the cost of the skateboard
   (b) the cost of the helmet

2. $Letting\ n\ represent\ the\ number$. Write each phrase as an algebraic expression.
   (a) The number increased by six
   (b) Three times the difference between the number and seven

3. Adam, Dexter and Blake spent the day fishing. Adam caught f fish. Dexter caught four times as many fish as Adam; while Blake caught seven less fish than Adam. Write down an expression in terms of f for
   (i) the number of fish caught by Dexter
   (ii) the number of fish caught by Blake

4. Johnathan bought a racing bike, gloves and helmet. The gloves cost $g\ dollars$. The bike cost ninety-five times as much as the gloves. The helmet cost \$34 more than the gloves. Write an expression, in terms of g for
   (a) the cost of the bike
   (b) the cost of the helmet

5. Koby bought baseball equipment for the summer games. The bat costs $b\ dollars$. The glove costs a third times the cost of the bat. The ball cost $12 less than the bat. Write an expression in terms of b for
   (a) the cost of the glove
   (b) the cost of the ball

6. $Letting\ k\ represent\ the\ number$. Write each phrase as an algebraic expression.
   (a) The number decreased by 7
   (b) The quantity 15 plus the number divided by 4

Substitution

Substitution means to replace. Therefore, when undergoing substitution in algebra we are required to replace the given number for each symbol in the algebraic expression. This replacement is to simplify and determine the particular numerical value of the given expression.

Note: When substituting, the following must be adhered

(1) Do not change the format of the expression
(2) Replace the symbol by the number directly into the algebraic expression to obtain an arithmetic expression and simplify
(3) When replacing the symbol with the given numerical value, it is best to place the given numerical value in a parenthesis.

Example: $Given\ x = 2;\ \ y = \ -3\ and\ z = 9, evaluate\ the\ following:$

(1) $x + y$

Working: $(2) + (-3)$

$= 2 - 3 = \ -1$

(2) $5x + y$

Working: $5(2) + (-3)$

$= 10 - 3 = \mathbf{7}$

(3) $\frac{2x-y}{z}$

Working: $\frac{2(2)-(-3)}{9}$

$= \frac{4\ +\ 3}{9}$

$= \frac{7}{9}$

(4) $x^3z\ +\ y$

Working: $(2)^3(9) + (-3)$

$= (8)(9) - 3$

$= 72 - 3 = \mathbf{69}$

(5) $\frac{xy^2 - 3x}{2z}$

Working: $\frac{(2)(-3)^2 - 3(2)}{2(9)}$

$= \frac{(2)(9) - 6}{18}$

$= \frac{18 - 6}{18} = \frac{12}{18}$; *which can be reduced to* $\frac{2}{3}$

$= \frac{2}{3}$

(6) Find the value of A from the formula $A = 2s^2 + 4sh; when\ s = 7\ and\ h = 3.$

Working:

Note: This formula is asking to find the value of A when given the other numerical values for s and h.

Therefore:

$$A = 2(7)^2 + 4(7)(3)$$

$$A = 2(49) + 84$$

$$A = 98 + 84$$

$$\boldsymbol{A} = 182$$

Activity 2a

(1) $Given\ that\ e = 2\ and\ g = -3, calculate\ the\ value\ of\ g + 4e$

(2) $Given\ that\ a = 5\ and\ c = -2, calculate\ the\ value\ of\ a^3 + 3c.$

(3) $Given\ that\ c = 6\ and\ f = 12, calculate\ the\ value\ of\ 3f - c^2.$

(4) $Given\ the\ value\ of\ b = 7\ and\ c = 0, calculate\ the\ value\ of\ b^2 - 4c.$

(5) $Given\ that\ x = 5\ and\ y = -4, evaluate\ x^2 + 3y.$

(6) $Given\ that\ s = 7\ and\ t = 12, calculate\ the\ value\ of\ s^2 - 4t.$

(7) $Given\ that\ p = 4\ and\ q = 10, evaluate:$

$(a)\ \frac{q}{p}$

$(b)\ 3p^2 + q^2$

(8) $Given\ that\ z = 9\ and\ y = 16, calculate\ the\ value\ of\ z^2 + \frac{y}{4}.$

Activity 2b

(1) For the formula K = P + Prt, find the value of K when P = 750, r = 0.05 and t = 8.

(2) A formula for conversion between temperature is:

$$C = \frac{5}{9}(F - 32)$$

Calculate C when F = 77

(3) $A\ formula\ for\ acceleration\ is$:

$$a = \frac{v^2 - u^2}{2x}$$

Find the value of a $when\ x = 14, v = 12\ and\ u = 8$, giving your answer correct to one decimal place.

(4) $A\ formula\ for\ the\ area\ of\ a\ trapezium\ is$:

$$\boldsymbol{A} = \frac{h(a + b)}{2}$$

Calculate the value of $\boldsymbol{A}$ when $h = 6.5, a = 8.4\ and\ b = 13.6$.

(5) A formula for velocity is, $v = \sqrt{2ax + u^2}$, find the value of $v\ when\ a = 4.5$, $x = 40$ and $u = 9$.

(6) The formula for the cost of electricity per month $b, in\ dollars$, for the use of E kilowatt-hours of electricity is

$$b = 40 + \frac{7E}{50}$$

Calculate the monthly bill when 605 kilowatt-hours are used.

(7) In house construction, the safe load $m\ kilograms$, that can be supported by a beam with length $x\ metres, thickness\ t\ centimetres, and\ height\ h\ centimetres$ is given by the formula:

$$m = \frac{4th^2}{x}$$

Calculate $m\ when\ t = 4.5, h = 20\ and\ x = 3.6$.

(8) The formula for the area of the walls of a regular room, ignoring doors and windows is:

$$A = 2h(l + w)$$

Calculate the area $A\ when\ h = 1.7metres, w = 4.3metres\ and\ l = 5.8metres.$

Algebraic Expressions

Adding and Subtracting Algebraic Terms:

Note: Algebraic terms can only be added to, or subtracted from each other if the terms are considered to be Like Terms.

- **Like Terms** – are those terms which are represented by the same algebraic symbols (variables) regardless of the size (magnitude) of each coefficient. However, the power (degree) of the symbols (variables) MUST be the same. Therefore, Like Terms are recognized by the same symbol/s with the same power/s.
- Additionally, ALL constants are Like Terms by way of being members of the Number System.

Note A: $5y\,,-3y,\ -0.8y,\ \frac{1}{2}y\ and\ -\frac{2}{5}y$ are **Like Terms** despite different coefficients because each variable is $y\ with\ the\ same\ power\ of\ 1\ for\ the\ y$. Therefore, all the terms can be added to, or subtract from each other.

Note: $5y\ is\ the\ same\ as\ 5y^1$.

Note B: $4n^2,\ n^2,\ -3n^2,\ 5n^3,\ 8p^2,\ 0.5n^2,$. All the terms given are Like Terms except $5n^3\ and\ 8p^2$. This is because, although the variables used for majority are $n, the\ degree\ (power) for\ four\ of\ the$ terms are raised to the 2nd power n^2 while $5n^3$ is raised to the 3rd power n^3. While $8p^2$ is raised to 2nd power, the variable is not the same as the others. As such $8p^2\ and\ 5n^3\ are\ referred\ to\ as\ \textbf{Unlike Terms}$. So, $4n^2,\ n^2,\ -3n^2,\ and\ 0.5n^2$ can be added to or subtract from each other, while $5n^3\ and\ \ 8p^2$ will stand alone, having nothing to add to, or subtract from.

Example: Simplify the following algebraic expressions

1. $Simplify \quad 7x + x + 12x$

   Note: All symbols do have a coefficient, if not seen then put a 1. So, we have:

   Working:

   $$7x + 1x + 12x = 20x$$

   Note: We ADD or SUBTRACT the coefficients based on the instructions given and keep the variable.

2. *Simplify* $15p - 9p + 4p$

Working:

$$15p - 9p + 4p = 10p$$

3. *Simplify* $9x^2 + 2x - 17x^2 + 18x$

Working:
Firstly, we group the Like Terms beside each other. Make sure to carry the signs the coefficients had before them

$$= 9x^2 - 17x^2 + 2x + 18x$$
$$= -8x^2 + 20x$$

4. *Simplify* $4a^2b - 7ab^2 + a^2b^2 - 3ab^2 + 3a^2b$

Working:
$= 4a^2b + 3a^2b - 7ab^2 - 3ab^2 + 1a^2b^2$
$= 7a^2b - 10ab^2 + 1a^2b^2$

5. A beverage set consists of a mug, a coffee pot and a serving plate. The mug costs p dollars. The serving plate costs $7 more than the mug. The coffee pot cost four times as much as the mug.

Expressing your answers in terms of p
(i) The cost of the serving plate
(ii) The cost of the coffee pot
(iii) The total cost of the set in simplest form

Answers:

(i) $p + 7$
(ii) $4 \times p = 4p$

(iii) **Note:** The beverage set consists of the mug $\boldsymbol{p}$, the serving plate $\boldsymbol{p+7}$ and the coffee pot $\boldsymbol{4p}$. So, the total cost of the set is

$$= p + p + 7 + 4p$$
$$= p + p + 4p + 7$$
$$= \mathbf{6}\boldsymbol{p} + \mathbf{7}$$

Activity 3

Simplify the following algebraic expressions

1. $3n + 7n$
2. $8x + 3x + x$
3. $7x - 2x$
4. $10x - 15x$
5. $-6p + 4p$
6. $-8r + 17r$
7. $15x - 9x + x$
8. $5n^2 + 7n^2$
9. $18x + 4p + 2x - 6p$
10. $13r^2 + 14r - 15r^2 - 12r$
11. $10x^2 - 9y^2 + 5y^2 - 7x^2$
12. $16p + 3 - 14p - 12$
13. $16a^2b + 4ab^2 - 3a^2b - 10ab^2 - 5$
14. $\frac{1}{4}y + \frac{1}{2}x + \frac{1}{8}y - \frac{3}{4}x$
15. $1.4xy - 3x^2y - 2.4xy^2 + 3.6xy - 7x^2y + 1.3xy^2$
16. $3gh + 8h + gh - 9h$
17. $2ab - 7a + 8a + 3ab$

Multiplying and Dividing Algebraic Terms:

Note: Commutative law holds true for multiplication of real numbers and algebraic terms. That is: $(i)\ 3 \times 5 = 5 \times 3 = 15\ \ (ii)\ a \times b = b \times a = ab\ or\ ba$; however, we tend to write in alphabetical order.

Multiplication

Example: Simplify each of the given algebraic expressions

1. $3x \times 5y$

Working:

$= 3 \times 5(x \times y)$
$= 15xy$

2. $\frac{1}{2}a \times 10b$

Working:

$= \frac{1}{2} \times 10(a \times b)$
$= 5ab$

3. $5p \times -3y$

Working:

$= - + 5 \times 3(p \times y)$
$= -15py$

4. $4a \times 2ab \times 3b \times \frac{3}{2}a^2b$

Working:

$= 4 \times 2 \times 3 \times \frac{3}{2}(a \times a \times a^2 \times b \times b \times b)$
$= 36(a \times a \times a \times a \times b \times b \times b) = 36a^4b^3$

5. $-7n \times -5 \times 4n$

Working:

$= -7 \times -5 \times 4(n \times n)$

$= - - 7 \times 5 \times 4(n \times n) = 140n^2$

6. $(3a^2b^3)(5ab^2)$

Above can be rewritten as: $3a^2b^3 \times 5ab^2$

Working:

$= 3 \times a \times a \times b \times b \times b \times 5 \times a \times b \times b$

$= 3 \times 5(a \times a \times a \times b \times b \times b \times b \times b)$

$= 15a^3b^5$

Division

Example: Simplify each of the given algebraic expressions

1. $x^5 \div x^3$

Note: **Division can rewrite as a fraction, then simplify**

Working:

$\Rightarrow x^5 \div x^3 = \frac{x \times x \times x \times x \times x}{x \times x \times x}$

$\Rightarrow \frac{1 \times 1 \times 1 \times x \times x}{1 \times 1 \times 1} = \frac{x^2}{1}$

$= x^2$

2. $8y^2 \div 4y^2$

Working:

$\Rightarrow \; 8y^2 \div 4y^2 = \frac{8 \times y \times y}{4 \times y \times y}$

$\Rightarrow \frac{8 \times 1 \times 1}{4 \times 1 \times 1} = \frac{8}{4} = \frac{2}{1}$

$= 2$

3. $12a^3b^3c \div 8ab$

Working:

$\Rightarrow \frac{12 \times a \times a \times a \times b \times b \times b \times c}{8 \times a \times b}$

$\Rightarrow \frac{12 \times 1 \times a \times a \times 1 \times b \times b \times c}{8 \times 1 \times 1}$

$\Rightarrow \frac{12a^2b^2c}{8}$

$= \frac{3a^2b^2c}{2}$ or $\frac{3}{2}a^2b^2c$

4. $-10y^2b^5 \div -5y^4b$

Working:

$\Rightarrow \; - - \frac{10 \times y \times y \times b \times b \times b \times b \times b}{5 \times y \times y \times y \times y \times b}$

$\Rightarrow \frac{10 \times 1 \times 1 \times 1 \times b \times b \times b \times b}{5 \times 1 \times 1 \times y \times y \times 1}$

$\Rightarrow \frac{10b^4}{5y^2}$

$= \frac{2b^4}{y^2}$

Activity 4

Simplify the following:

1. $a^2 \times a^4$
2. $3m^4 \times 2m \times m^3$
3. $8x^2y \times 2x^2y^5$
4. $0.7ab^2c \times 0.9a^2b^2c$
5. $\frac{2}{3}m^2n^2 \times \frac{3}{4}mn^2$
6. $\frac{4}{5}x^2y^2 \times 10x^3y$
7. $7p^2q \times 2 \times 3pq$
8. $12y^9 \div 4y^5$
9. $3x^5 \div 15x^4$
10. $16x^4y^8 \div 2x^2y^2$
11. $6a^2b^5 \div 4a^2b^3c$
12. $-150m^7n^2 \div -10m^3$
13. $-0.5p^3q^4 \div 0.02p^4q^2$
14. $\frac{20a^4}{10a}$
15. $\frac{8m^4n^3}{12m^2n^2}$
16. $\frac{12r^3y^2z}{4ry}$

Removing Brackets

Note: In order to remove brackets from an algebraic expression, we MUST apply the *Distributive Law*. This law requires the multiplication of any term outside the bracket by each term inside the bracket. As such, $(a + b)x$ *which can be rewritten as* $x(a + b)$; and when applying the Distributive Law, we see that:

$$(a + b)x = x(a + b) = x(a) + x(b) = ax + bx$$

Example:

(A) Remove the brackets (or expand) the following:

1. $8(x + y)$

Working:

$\Rightarrow 8(x) + 8(+y)$

$= 8x + 8y$

2. $3(ab - c)$

Working:

$\Rightarrow 3(ab) + 3(-c) = 3(ab) - +3(c)$

$= 3ab - 3c$

3. $-5(x + 2y)$

Working:

$\Rightarrow -5(x) - 5(+2y)$

4. $-2(3a - 4b)$

Working:

$\Rightarrow -2(3a) - 2(-4b)$

$= -5x - 10y$

$= -6a + 8b$

5. $n(3a - n)$

Working:

$\Rightarrow n(3a) + n(-n) = 3an - n^2$

6. $-(2a - b)$

Working: Rewritten as: $-1(2a - b)$

$\Rightarrow -1(2a) - 1(-b) = -2a + 1b$

7. $8\left(x - \frac{1}{2}\right)$

Working: $8(x) + 8\left(-\frac{1}{2}\right) = 8x - 4$

(B) Remove the brackets and simplify

1. $8(x + y) - 5(x - 2y)$

Working:

$\Rightarrow 8(x) + 8(+y) - 5(x) - 5(-2y)$

$\Rightarrow 8x + 8y - 5x + 10y$

$\Rightarrow 8x - 5x + 8y + 10y$

$= 3x + 18y$

2. $2 - 5(y - 7) - 3y$

Working: Removing brackets first

$\Rightarrow 2 - 5(y) - 5(-7) - 3y$

$\Rightarrow 2 - 5y + 35 - 3y$

$\Rightarrow 2 + 35 - 5y - 3y$

$= 37 - 8y$

3. $6(m + n) - 5(m - 3n)$

Working:

$\Rightarrow 6(m) + 6(+n) - 5(m) - 5(-3n)$

$\Rightarrow 6m + 6n - 5m + 15n$

$\Rightarrow 6m - 5m + 6n + 15n$

$= 1m + 21n$

4. $\frac{1}{2}(4x + 6) + \frac{2}{5}(10x + 5)$

Working:

$\Rightarrow \frac{1}{2}(4x) + \frac{1}{2}(+6) + \frac{2}{5}(10x) + \frac{2}{5}(+5)$

$\Rightarrow 2x + 3 + 4x + 2$

$\Rightarrow 2x + 4x + 3 + 2$

$= 6x + 5$

Activity 5a

Remove the brackets

1. $7(3x + y)$
2. $-4(a + 2b)$
3. $-2(n - 2m)$
4. $6\left(\frac{1}{3} - x\right)$
5. $-(a + 2b)$
6. $\frac{1}{3}(6a + 3)$
7. $0.5(2a - 8b)$
8. $3\left(\frac{1}{2} - 2x\right)$

Activity 5b

Expand and simplify

1. $4(a+b)+2(a+3b)$ 2. $5(x-y)-3(x+y)$ 3. $8(n+m)-(n-5m)$

4. $2(x+2y)-3(x-4y)$ 5. $\frac{1}{4}(x+8)+\frac{3}{4}(x+4)$ 6. $\frac{1}{2}(4+2y)+3(2+y)$

7. $\frac{2}{5}(5+10y)-\frac{1}{3}(3-6y)$ 8. $7+4(x-5)-3x$ 9. $8-3(2x-5)+7x$

Factorizing Expressions

Note: When we factorize algebraic expressions, we are expressing them as a product of some of their factors. When we use the Distributive Law to insert brackets in an expression, we are factorizing.

So: $ax+bx=x(a+b)$

Where x *is the common factor for ax and bx.*

Note: we take out the common factor x, *and place the other factor* $(a+b)$ in bracket.

Example:

Factorize completely

1. $4a+4d$

Working: H.C.F of 4 & 4 is **4**

Breaking down each term into products

$\Rightarrow \cancel{4}.a+\cancel{4}.d$

Note: 4 is the H.C.F. for the two terms

$Answer \ = \ 4(a+d)$

2. $5c-5b$

Working: H.C.F. of 5 and 5 is **5**

Breaking down each term into products

$\Rightarrow \cancel{5}.c-\cancel{5}.b$

Note: 5 is the H.C.F. for the two terms

$Answer \ = 5(c-b)$

3. $25c+10$

Working: the H.C.F. of 25 & 10 is 5

Breaking down each term into products

So $\Rightarrow \cancel{5}.5.c+\cancel{5}.2$

Note: 5 is the H.C.F. for the two terms

$Answer = 5(5c+2)$

4. $8a^2-64a$

Working: the H.C.F. of 8 & 64 is 8

Breaking down each term into products

So $\Rightarrow \cancel{8}.1.\cancel{a}.a-\cancel{8}.8.\cancel{a}$

Note: 8a is the H.C.F. for the two terms

$Answer = 8a(1a-8)$

5. $6x^3y - 8xy^2 + 12xy$

Working: the H.C.F. of 6, 8 & 12 is 2

Breaking down each term into products

$\not{2}.3.\not{x}.x.x.\not{y} - \not{2}.4.\not{x}.\not{y}.y + \not{2}.6.\not{x}.\not{y}$

Note: $2xy$ is the H.C.F. for the three terms

Answer $= 2xy(3x^2 - 4y + 6)$

6. $12mn^3 + 9m^2n^2$

Working: the H.C.F, of 12 & 9 is 3

Breaking down each term into products

$\not{3}.4.\not{m}.\not{n}.\not{n}.n + \not{3}.3.\not{m}.m.\not{n}.\not{n}$

Note: $3mn^2$ *is the H.C.F. for the two terms*

Answer $= 3mn^2(4n + 3m)$

Activity 6a

Factorize completely

1. $2a + 4$
2. $mx - nx$
3. $18a - 6c$
4. $12x - 16$
5. $3ab + 6abc$

Activity 6b

Factorize completely

1. $27x^2y - 9xy$
2. $6a^3b + 3ab^2$
3. $15a^3 - 30a^2$
4. $7m^2n - 21mn$
5. $18a^2b - 12ab^2$
6. $12ab^2 + 18a^2b - 6b$
7. $3a^2b - 6ab + 5ab^2$
8. $4mn^2 + mn + 3m^2n$
9. $35y^3 - 7y + 14y^2$
10. $12x^2y + 8xy^2 - 4y$

Addition and Subtraction of Algebraic Fractions

Using a constant as denominator:

Note: We use the same concepts when adding or subtracting fractions; which is to find the L.C.D. (lowest common denominator).

Most of the times when adding or subtracting fractions the answer remains as a fraction or it may simplify to a whole.

Example:

Simplify

1. $\frac{4y}{5} + \frac{2y}{3}$

Note: *The L.C.D. of 5 & 3 is 15. So the aim is to make the denominators 5 & 3 to become 15. Hence, we multiply the 5 by 3 and the 3 by 5.*

Since we adjusted the denominators, we have to adjust the numerators 4y and 2y by multiplyimg them by the same numbers used to change the denominators to the required L.C.D.

Thus: $\frac{3(4y)}{3(5)} + \frac{5(2y)}{5(3)} = \frac{3(4y)+5(2y)}{15}$

$\Rightarrow \frac{12y+10y}{15} = \frac{22y}{15}$ which simplify to give $1\frac{7y}{15}$ *We keep the single denominator and add the numerator.*

Simplify

2. $\frac{5x}{6} - \frac{3x}{4}$

Note: *The L.C.D. of 6 & 4 is 12. And using the same concepts as Example 1.*

To get 6 and 4 to 12 we multiply by 2 and 3 respectively.

Working:

$= \frac{2(5x)}{2(6)} - \frac{3(3x)}{3(4)} = \frac{2(5x)-3(3x)}{12} = \frac{10x-9x}{12} = \frac{1x}{12}$

Simplify

3. $\frac{3}{7}n + \frac{1}{14}n$

The above algebraic fractional expression can be rewritten as $\frac{3n}{7} + \frac{1n}{14}$

The L.C.D. of 7 and 14 is 14

Working:

$= \frac{2(3n)}{2(7)} + \frac{1(1n)}{1(14)} = \frac{2(3n)+1(1n)}{14} = \frac{6n+1n}{14} = \frac{7n}{14}$

The answer above can be reduced to give $= \frac{1n}{2}$

Activity 7

Simplify the following:

1. $\frac{3y}{4} - \frac{2y}{5}$ 2. $\frac{3a}{5} + \frac{a}{6}$ 3. $\frac{x}{2} - \frac{x}{6}$ 4. $\frac{3xy}{10} + \frac{2xy}{5}$

5. $\frac{7}{8}y - \frac{3}{4}y$ 6. $\frac{2m}{5} + \frac{1}{3}m$ 7. $\frac{9}{10}x + \frac{2}{5}x$ 8. $\frac{2}{3}y - \frac{8}{5}y$

Solving Linear Equations

Left-hand-side Right-hand-side

Note: A linear equation can have only ONE solution. $(L.H.S.) = (R.H.S.)$

[The equal sign separates what is on the L.H.S. to what is on the R.H.S.] $x + 8 = 15$

An equation can be likened to a double beam balance:

The solution for the above equation will be $x = 7$. *Since* $7 + 8 = 15$; *which balance both sides of the double beam balance making it 15 = 15.*

We can maintain balance or equality in an equation by:

- Adding or subtracting the inverse quantity to both sides of the equation based on the given operation.

- Multiply or divide by the inverse quantity to both sides of the equation based on the given operation.

Solving One-Step Equation:

Example:

Solve

1. $x + 5 = 3$

Working: *The aim is to find positive x, so we need to to do the inverse of +5 which is – 5 to both sides*

$\Rightarrow x + 5 - 5 = 3 - 5$

$\Rightarrow x = -2$

Alternative Method

$x + 5 = 3$

Working:

$x = 3 - 5$

$x = -2$

We remove the +5 located on the L.H.S. to the R.H.S.
The +5 changes to – 5 as it is now relocated to the R.H.S.

Solve

2. $x - 8 = 4$

Working: *The aim is to find positive x, so we need do the inverse of – 8 which is +8 to both sides*

$\Rightarrow x - 8 + 8 = 4 + 8$

$\Rightarrow x = 12$

Alternative Method

$x - 8 = 4$

Working:

$x = 4 + 8$

$x = 12$

We remove the – 8 located on the L.H.S. to the R.H.S.
The – 8 changes to +8 as it is now relocated to the R.H.S.

3. Solve

$2x = 5$

Working:

$x = \frac{5}{2} = 2.5$

To find x we need to get rid of the coefficient 2 by dividing Itself out and then dividing it into the 5 on the R.H.S.

4. Solve

$\frac{3}{2}x = 6$

Working:

$x = 6 \div \frac{3}{2} \quad or \quad x = \frac{6}{\frac{3}{2}}$

$x = 4$

To find x we need to get rid of the coefficient $\frac{3}{2}$ by Dividing itself out and then dividing into the 6 on the R.H.S.

5. Solve

$2 + x = 7$

Working:

Note: the 2 is positive

$So,\ +x = 7 - 2$

$+x = 5$

$x = 5$

6. Solve

$0.5x = 2$

Working:

$x = \frac{2}{0.5} = 4$

$x = 4$

To find x we need to get rid of the coefficient 0.5 by dividing itself out and then divide into 2 on the R.H.S.

7. $Solve\ \frac{x}{3} = 5$

Working: *To remove the division by 3, we remove it by multiplying by 3 on the R.H.S.*

$x = 5(3)$
$x = 15$

Solving Two-Steps Equation:

1. $Solve\ the\ equation$

$5x - 3 = 15$

Working: $1st\ step; relocating\ the\ -3\ to\ the$ R.H.S.

$\Rightarrow 5x = 15 + 3$ $[Adding\ 15 + 3]$

$\Rightarrow 5x = 18$

$2nd\ step; dividing\ out\ the\ coefficient\ 5$

$\Rightarrow x = \frac{18}{5} = 3.6$

$x = 3.6$

2. $Solve\ the\ equation$

$7 + 2x = 19$

Working: $Relocating\ +7\ to\ the\ R.H.S$

$\Rightarrow +2x = 19 - 7$ $[Subtracting\ 19\ \ 7]$

$\Rightarrow +2x = 12$

$2nd\ step; dividing\ out\ the\ coefficient\ 2$

$\Rightarrow x = \frac{12}{2} = 6$

$x = 6$

3. $Solve$

$2 - x = 5$

$\boldsymbol{The\ above\ equation\ can\ be\ rewritten\ as\ 2 - 1x = 5}$

Working: $1st\ step; relocating\ +2$ to R.H.S.

$\Rightarrow -1x = 5 - 2$

$\Rightarrow -1x = 3$

$The\ variable\ x\ cannot\ be\ negative. So\ we\ need\ to\ divide\ out\ by\ -1\ coefficient.$

$\Rightarrow x = \frac{3}{-1} = -3$

$x = -3$

4. $Solve$

$3 - 4x = 15$

Working: 1st step; relocating the +3 to R.H.S.

$\Rightarrow -4x = 15 - 3$

$\Rightarrow -4x = 12$

$The\ variable\ x\ cannot\ be\ negative. So\ we$ need to divide out by $-4\ coefficient.$

$\Rightarrow x = \frac{12}{-4} = -3$

$x = -3$

Solving Equations Containing The Unknown On Both Sides:

Example:

Solve the equation $5x - 4 = 3x + 16$

Working: *Relocating the Like Terms together on a side. In this case, the choice is to relocate the terms with x on the L.H.S. and relocate the constants on the R.H.S.*

$\Rightarrow 5x - 3x = +16 + 4$

From above: $The -4\ on\ the\ L.H.S. when\ relocated\ to\ the\ R.H.S\ became\ +4; while\ the\ +3x\ on\ the\ R.H.S. when\ relocated\ to\ the\ L.H.S. became\ -3x.$

$\Rightarrow 2x = 20$

$\Rightarrow x = \dfrac{20}{2} = 10$

$x = 10$

Equation Containing Brackets:

Note: When an equation contains bracket/s, we firstly use the Distributive Law to remove the bracket/s then solve.

Example:

Solve the equation $2(3x - 5) = 14$

Working:

$\Rightarrow 2(3x) + 2(-5) = 14$

$\Rightarrow 6x - 10 = 14$ **Note:** ***A two step equation is created***

$\Rightarrow 6x = 14 + 10$

$\Rightarrow 6x = 24$

$\Rightarrow x = \dfrac{24}{6} = 4$

$x = 4$

Activity 8a

Solve each of the following equations:

1. $x + 7 = 15$
2. $x - 5 = 0$
3. $x + 5 = -1$
4. $5x = 12$
5. $0.4x = 16$
6. $3 + x = 7$
7. $\frac{2}{5}x = 6$
8. $\frac{x}{9} = 4$
9. $\frac{x}{8} = \frac{7}{4}$
10. $\frac{x}{5} = \frac{4}{10}$
11. $2x + 7 = 17$
12. $2 - x = -3$
13. $3x + 1 = 4$
14. $0.2x + 3 = 1.7$
15. $\frac{x}{7} - 4 = 0$
16. $\frac{x}{2} + 5 = 9$

Activity 8b

Solve

1. $7x - 3 = 4x + 12$
2. $3x + 4 = 2x + 7$
3. $6x - 15 = 4x - 7$
4. $5 + 4x = 13 + 2x$
5. $3 - 3x = 17 - 5x$
6. $1 + x = 7 - 2x$
7. $2(x - 7) = 0$
8. $3(1 + 2y) = 27$
9. $5\left(x - \frac{2}{5}\right) = 12$
10. $\frac{1}{3}(6x + 12) = 16$

Solving Algebraic Fractional Equations:

Example:

1. Solve $\frac{x}{5} - \frac{2}{7} = \frac{4}{5}$

Working: *Relocating the $-\frac{2}{7}$ on R.H.S*

Simplifying the two constants $\frac{4}{5}$ and $\frac{2}{7}$.

$\Rightarrow \frac{x}{5} = \frac{4}{5} + \frac{2}{7}$

$\Rightarrow \frac{x}{5} = \frac{7(4)}{7(5)} + \frac{5(2)}{5(7)}$ *L.C.D is 35*

$\Rightarrow \frac{x}{5} = \frac{28+10}{35}$

$\Rightarrow \frac{x}{5} = \frac{38}{35}$

Removing the 5 which is being divided into the x

by multiplying it by the $\frac{38}{35}$ on the R.H.S.

$\Rightarrow x = \frac{38}{35} \times 5 = \frac{38}{7}$

$\Rightarrow x = \frac{38}{7} = 5\frac{3}{7}$

$x = 5\frac{3}{7}$

3. $\frac{2x}{5} = 9 - \frac{x}{2}$

Working: *Relocating $-\frac{x}{2}$ to the L.H.S. with it's Like Term*

$\Rightarrow \frac{2x}{5} + \frac{x}{2} = 9$

$\Rightarrow \frac{2(2x)}{2(5)} + \frac{5(x)}{5(2)} = 9$ *L.C.D. Is 10*

$\Rightarrow \frac{2(2x)+5(x)}{10} = 9$

$\Rightarrow \frac{2(2x)+5(x)}{10} = 9$

$\Rightarrow \frac{4x+5x}{10} = 9 \Rightarrow \frac{9x}{10} = 9$

Relocating the coefficient $\frac{9}{10}$ to divide it inot 9 on the R.H.S.

$\Rightarrow x = 9 \div \frac{9}{10}$

$x = 10$

2. *Solve* $\frac{5x}{7} - \frac{x}{14} = \frac{9}{7}$

Working:

They can be subtracted because they are Like Terms.

$\Rightarrow \frac{2(5x)}{2(7)} - \frac{1(x)}{1(14)} = \frac{9}{7} \Rightarrow \frac{2(5x)-1(x)}{14} = \frac{9}{7}$

$\Rightarrow \frac{10x-1x}{14} = \frac{9}{7}$ *L.C.D is 14*

$\Rightarrow \frac{9x}{14} = \frac{9}{7}$

Removing $\frac{9}{14}$ from being a coefficient of x by

diving by $\frac{9}{14}$ *into the* $\frac{9}{7}$ on the RH.S.

$\Rightarrow x = \frac{9}{7} \div \frac{9}{14}$ $\Rightarrow x = 2$

4. $\frac{n-3}{2} = \frac{n}{5}$

Rewriting the above as $\frac{n}{2} - \frac{3}{2} = \frac{n}{5}$; then

Relocating $\frac{3}{2}$ to the R.H.S. and the $\frac{n}{5}$ to the L.H.S.

Working: $\frac{n}{2} - \frac{n}{5} = \frac{3}{2}$ *L.C.D is 10*

$\Rightarrow \frac{5(n)}{5(2)} - \frac{2(n)}{2(5)} = \frac{3}{2}$

$\Rightarrow \frac{5(n)-2(n)}{10} = \frac{3}{2} \Rightarrow \frac{5n-2n}{10} = \frac{3}{2}$

$\Rightarrow \frac{3n}{10} = \frac{3}{2}$

Removing $\frac{3}{10}$ from being a coefficient of n by

dividing by $\frac{3}{10}$ *into the* $\frac{3}{2}$ on the RH.S.

$\Rightarrow n = \frac{3}{2} \div \frac{3}{10}$

$n = 5$

- **Please be note that the operation of a coefficient is MULTIPLICATION. Therefore, whenever we relocated a coefficient to solve an equation, then we apply the inverse operation which is to DIVIDE.**

Activity 9

Solve

1. $\frac{y}{2} - \frac{4y}{5} = \frac{9}{10}$
2. $\frac{y}{10} = 9 - \frac{y}{5}$
3. $\frac{x}{2} + \frac{5}{6} = \frac{x}{3}$
4. $\frac{x}{2} - \frac{x}{6} = \frac{3}{4}$
5. $\frac{3x}{5} - \frac{1}{4} = \frac{3x}{10}$
6. $\frac{2c-5}{7} = \frac{c}{3}$
7. $\frac{3x-1}{5} = \frac{2x+7}{10}$

Inequalities

The four inequality symbols are: $<$, $>$, $\leq$, $\geq$

$x < a$ *means x is less than a*	$x > a$ *means x is greater than a*
$x \leq a$ *means x is less than or equal to a*	$x \geq a$ *means x is greater or equal to a*

Solving Linear Inequation with One Unknown:

The inequality signs are: $<, \leq, >$ *and* $\geq$ *and are used to represent inequations.* An inequation is a statement involving an inequality. It shows a statement that one quantity is not equal to another quantity. A linear inequation with one unknown is an inequation where the solution is a range of values and hence, it is given as a solution set. For example: $4x + 7 > 7$, *where the solution has a range of values.*

Note:

To solve an inequation, we use the same concepts as solving a linear equation; except we keep the inequality symbol ($<$ or $>$ or $\leq$ or $\geq$) given in the inequation.

Solving inequations, where the given coefficient is NEGATIVE

Whenever we used to NEGATIVE coefficient to Divide or Multiply, to determine the solution set, the inequality sign MUST be REVERSED

Example: 1

$-ax > b$ where $-a$ *and* b *are constants*

Then $x < \frac{b}{-a}$ *As seen the inequality sign Reverses*

Example: 2

$-\frac{x}{a} \leq b$

Then $x \geq b(-a)$ *As seen the inequality sign Reverses*

$x \geq -ab$

Example:

1. (a) *Solve the inequation* $x + 3 \geq 7$

(b) *State the solution set*

(c) Represent the solution set on a directed number line

Working:

(a) $x + 3 \geq 7$

$x \geq 7 - 3$

$x \geq 4$

(b) *solution set* $= \{x : x \geq 4\}$

(c)

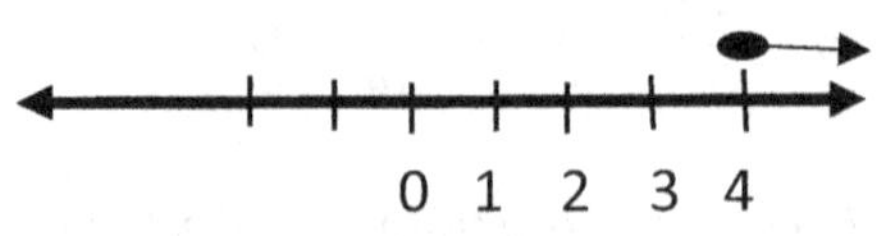

Note: On the number line

$> \& <$ *uses an open circle*

$\geq \& \leq$ *uses a closed circle*

The unshaded circle indicates that the unknown is not equal to, while the shaded circle indicates the unknown is equl to.

Given $x > b$ *or* $x \geq b$; *where b is a constant, procudes an arrow directed to the right (always read from the variable):*

→

Given $x < b$ *or* $x \leq b$; *where b is a constant, produces an arrow directed to the left (always read from the variable):*

2. (a) *Solve the inequation* $x + 3 < 2$

(b) *State the solution set*

(c) Represent the solution set on a directed number line

Working:

(a) $x + 3 < 2$ (b) *solution set* $= \{x: x < -1\}$ (c)

-2 -1 0 1 2

$x < 2 - 3$

$x < -1$

3. *Solve* $2x \leq 4$

Working:

$x \leq \frac{4}{2}$

$x \leq 2$

4. *Solve* $-2x \leq 4$

Working:

$x \geq \frac{4}{-2}$ *The direction of the inequality sign reverses.*

$x \geq -2$ *This is because we divide by* -2 *coefficient.*

5. *Solve* $-\frac{x}{2} > 3$

Working:

$x < 3(-2)$ *To make x positive we had to multiply*

$x < -6$ *by* -2 *coefficient on the R.H.S.*

So the inequality sign reverses.

Activity 10a

Express each of the following statement as an inequality statement

1. *x is less than* 19
2. *y is greater than* -1
3. *p is not more than* 25
4. *q is at least* 12

Activity 10b

Solve each of the following inequations (b) state the solution set (c) represent the solution set on a directed number

1. $x + 5 < 7$
2. $x + 9 \geq 4$
3. $5x \leq 45$
4. $2x > -2$
5. $-6x \geq 18$
6. $-0.2x \geq 1$
7. $\frac{x}{5} < 0.2$
8. $-\frac{2}{5}x < -0.4$

9. $2(x-1) \leq 10$ 10. $0.2(x-10) > 1$
11. $3x+2 \leq -5x-10$
12. $4(x+1) \leq 2x+6$

Simultaneous Equations

Simultaneous equations are a system of several equations with several unknowns. Simultaneous equations share variables. These equations are solved simultaneously (at the same time).
Simultaneous equations are equations that have the same solutions. This means, the equations are all satisfied by the same values of the unknown quantity.
Solving simultaneously algebraically may use two methods: (i) Elimination method (ii) Substitution method.

Elimination Method:

As suggested, the aim is to eliminate one variable (to make it become zero) firstly so that you can find the other variable.
In order to do so, we have to make the coefficients of the variables we wish to eliminate similar. This means the variables should have the same absolute value. Meaning, they have the inverse operation but the same constant.

Example:
$-5 \,\&+5$ ***Absolute value are the same. They both have a distance of*** 5 ***space away form zero*** (0).
However, $-5+5=0$ ***or*** $+5-5=0$.
So, if we decide to eliminate $-2y$ ***then we must have*** $+2y$***. Hence,*** $-2y+2y$ ***gives*** 0.

After we eliminate the first variable, we use the concepts of solving linear equations to find the other variable. We finally, use anyone of the previous equations and solve for the eliminated variable by substitution and the concepts of solving linear equations. When both coefficients have the same sign such as (+ & +) or (−& −); to change the sign of one of the coefficients you MUST multiply one of them by the required negative and the other by the required positive number.

Example:

1. Solve the pair of simultaneous equations

$3x + y = 10$

$2x - y = 15$

$\textbf{\textit{Note}}: \textit{Both } + y \textit{ and } - y \textit{ can elimante because the coeffient of the y's are Absolute value of eahc other; } +1 \textit{ and } - 1.$

Working:

$3x + 1y = 10$ **eqn. (1)**

$2x - 1y = 15$ **eqn. (2)**

$5x + 0 = 25$

$\Rightarrow 5x = 25$

$\Rightarrow x = \frac{25}{5}$

$x = 5$

$\textbf{\textit{Finding y by using eqn.1 where x}} = \textbf{5}$

$3x + 1y = 10$

$3(5) + 1y = 10$

$15 + 1y = 10$

$1y = 10 - 15$

$1y = -5$

$y = -5$

$\textit{Hence, the solutions are}: \quad x = 5 \textit{ and } y = -5$

2. Solve the pair of simultaneous equations

$3x + 5y = 21$

$2x + 3y = 13$

Working:

$\textit{We are going to elimate the } + 5y \textit{ and } + 3y \textit{ variable. But both are positive so we need one of them to become negative.}$

$\textit{To do this; seeing 15 is a multiple of both coefficients.}$ To eliminate them:

$\textit{We will create the Absolut value } - 15y \textit{ and } + 15y \textit{ by multiplying eqn. 1 by } - 3 \textit{ and eqn. 2 by } + 5 \textit{ or vice vera.}$

$(3x + 5y = 21) \times -3 \quad \Rightarrow \quad -9x - 15y = -63$

$(2x + 3y = 13) \times 5 \quad \Rightarrow \quad 10x + 15y = 65$

$1x + 0 = 2$

$1x = 2$

$x = 2$

$\textbf{\textit{Finding y by using eqn.2 where x}} = \textbf{2}$

$2x + 3y = 13$

$2(2) + 3y = 13$

$4 + 3y = 13$

$3y = 13 - 4$

$3y = 9$

$y = \frac{9}{3}$

$y = 3$

$\textit{Hence, the solutions are}: \quad x = 2 \textit{ and } y = 3$

3. Solve the pair of simultaneous equations
$3x - 2y = 3$
$2x = 1 + y$

Working:

As seen above eqn. 2 is not arranged like eqn. 1 where the variables are on the L.H.S. and the constant on the R.H.S. Therefore, before we solve the equations, we will REARRANGE eqn. 2 to be arranged or lined up like eqn. 1. This now give:

$3x - 2y = 3$
$2x - y = 1$

Now: $-2y$ *and* $-y$ *are given, showing both coefficients are negative.*
To elimate the y we must have $+2y$ *and* $-2y$. *So we will multiply eqn.* 1 *by* -1 *and eqn.* 2 *by* 2.

$(3x - 2y = 3) \times -1 \quad \Rightarrow -3x + 2y = -3$
$(2x - y = 1) \times 2 \quad \Rightarrow \ \ 4x - 2y = \ \ 2$

$1x + 0 = -1$
$1x = -1$
$x = -1$

Finding y by using eqn. 2 where x = −1

$2x - y = 1$
$2(-1) - y = 1$
$-2 - y = 1$
$-y = 1 + 2$
$-y = 3$ which means $-1y = 3$

So, $y = \dfrac{3}{-1}$

$y = -3$

Hence, *the solutions are*: $x = -1$ *and* $y = -3$

4. Solve the pair of simultaneous equations
$3x - 2y = 7$
$-x + 3y = -7$

Working:

$-2y$ & $+3y$ *are given, showing one coefficient* $-$ *ve & the other* $+$ *ve.*
Since we already have a $+$ *ve and* $-$ *ve coefficient we just multiply eqn.* 1 *and eqn.* 2 *by positive* 3 *and* 2 *respectively. To get* $-6y$ *and* $+6y$ *so they can be eliminated.*

$(3x - 2y = 7) \times 3 \quad \Rightarrow 9x - 6y = \ \ 21 \quad \Rightarrow 7x = 7$
$(-x + 3y = -7) \times 2 \quad \Rightarrow -2x + 6y = -14 \quad \Rightarrow x = \frac{7}{7}$

$7x + 0 = 7 \quad x = 1$

Finding y by using eqn. 1 where x = 1

$3x - 2y = 7$

$3(1) - 2y = 7$
$3 - 2y = 7$
$-2y = 7 - 3$
$-2y = 4$
$y = \frac{4}{-2}$
$y = -2$

Hence, $the\ solutions\ are:\ \ x = 1\ and\ y = -2$

Substitution Method:

Example:

Solve the pair of simultaneous equations

$$5x + 2y = 29$$
$$x - y = -4$$

Working:

Taking equation (2) and making x the subject gives: $x = -4 + y$

Now, substituting for x into equation (1) gives:

$\Rightarrow 5(-4 + y) + 2y = 29$ *Removing the bracket using the Distributive law gives:*
$\Rightarrow -20 + 5y + 2y = 29$ *Gathering Like terms on each side gives:*
$\Rightarrow 5y + 2y = 29 + 20$
$\Rightarrow 7y = 49$
$\Rightarrow y = \frac{49}{7} = 7$

$Using\ equation\ 2\ to\ find\ the\ value\ of\ x; where\ y = 7$

$x - y = -4$
$x - (7) = -4; \quad \Rightarrow x - 7 = -4$
$x = -4 + 7$
$x = 3$

The solutions are: $x = 3\ \ and\ \ y = 7$

Word Problem – Simultaneous Equations:

Example:

Fred bought 3 breads and 5 buns for \$32.75. However, if he had bought 4 breads and 4 buns, he would have paid \$29.00.

(a) Write down a pair of simultaneous equations from the above information, using $x\ and\ y$ to represent one bread and one bun respectively.

(b) Solve your pair of simultaneous equations to find the cost of one bread and one bun.

Working:

(a) $3x + 5y = \$32.75$

$4x + 4y = \$29.00$

(b) $(3x + 5y = \$32.75) \times -4 \quad \Rightarrow -12x - 20y = -131$

$(4x + 4y = \$29.00) \times 5 \quad \Rightarrow 20x + 20y = 145$

$8x \ + \ 0 \ = \ 14$

$8x = 14$

$x = \frac{14}{8}$

$x = \$1.75$

Using eqn. 1 to find y, where $x = \$1.75$

$3x + 5y = \$32.75$

$3\ (\$1.75) + 5y = \$32.75$

$\$5.25 + 5y = \$32.75$

$5y = \$32.75 - \$5.25$

$5y = \$27.50$

$y = \frac{27.50}{5}$

$y = \$5.50$

$Hence, the\ cost\ of\ one\ bread(x) = \$1.75\ and\ one\ bun(y) = \$5.50$

Activity 11

Solve the pair of simultaneous equations

1. $3x + y = 3$
   $2x - 3y = 13$
2. $2x + 7y = 0$
   $x - 6y = 19$
3. $2x + y = 14$
   $-3y + 4x = 8$
4. $1 + 4x = 2y$
   $-4 + 3y = 8x$
5. $7x - 3y = 27$
   $y = -2x + 4$
6. $y = 4x + 1$
   $y = 4 + 2x$

7. $x - y = -5$
   $2y = 3x$

8. $y = -2x + 5$
   $3y + x = 0$

(9) $A\ rectangle\ has\ dimensions\ given\ in\ terms\ of\ x\ and\ y\ as\ shown.$

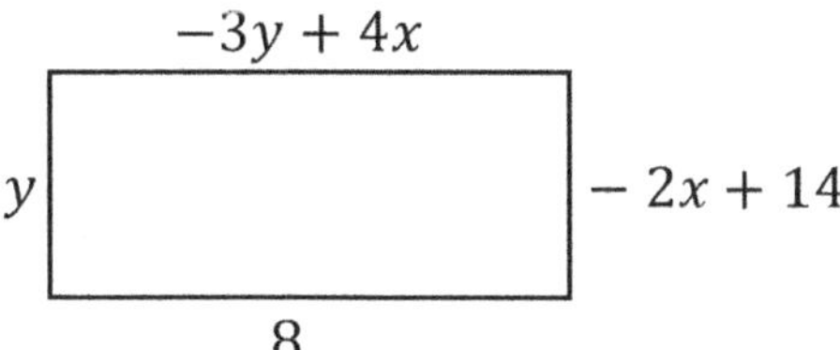

(a) $Write\ down\ a\ pair\ of\ simultaneous\ equations\ for\ the\ length\ and\ width\ of\ the\ above\ rectangle.$

$(b)\ Solve\ your\ simultaneous\ equations\ to\ find\ the\ values\ for\ x\ and\ y.$

10. $Ben\ has\ x\ dimes\ and\ y\ quarters. He\ has\ a\ total\ of\ 17\ coins\ with\ a\ combined\ value\ of\ \$3.05.$

$(a)\ Write\ an\ equation\ in\ terms\ of\ x\ and\ y\ that\ represents\ the\ number\ of\ coins.$

$(b)\ Write\ an\ expression\ for\ the\ value$

   $(i)\ \ the\ dime$

   $(ii)\ the\ quarter$

$(c)\ Write\ an\ expression, in\ terms\ of\ x\ and\ y\ that\ represents\ the\ combined\ value\ of\ the\ coins.$

$(d)\ Solve\ your\ system\ of\ equations\ [(a) and\ (c)], to\ find\ the\ actual\ number\ of\ quarters\ and\ dimes.$

Indices

Note: $a \times a \times a \times a \times a = \boldsymbol{a}^5$, $which\ is\ read\ as\ \boldsymbol{a}\ to\ the\ power\ 5. This\ format\ is\ known\ as\ INDEX\ FORM.$

$The\ \boldsymbol{a}\ is\ reffered\ to\ as\ the\ base\ and\ the\ number\ 5\ is\ reffered\ to\ as\ the\ index\ (power\ or\ exponent). This\ expression\ is\ saying\ the\ base\ \boldsymbol{a}\ is\ raised\ to\ the\ power\ (index\ or\ exponent)\ 5.$

$When\ an\ algebraic\ quantity\ or\ number\ is\ multiplied\ by\ itself\ repeatedly, then\ it\ can\ be\ expressed\ as\ a\ base\ raised\ to\ a\ power. Hence, a \times a \times a \times a \times \ldots to\ the\ nth\ term\ gives\ a^n, where\ a \neq 0.$

Example:

Express each of the following in index form:

1. $a \times a \times a \times a$
2. $5 \times 5 \times 5 \times 5 \times 5 \times 5$

Working:

1. $a \times a \times a \times a = a^4$
2. $5 \times 5 \times 5 \times 5 \times 5 \times 5 = 5^6$

Laws of Indices:

The laws of indices are summarized below:

1. **Product Rule states:**

   $a^m \times a^n = a^{m+n}$

Example:

Simplify

(a) $a^5 \times a^6 = a^{5+6} = a^{11}$

(b) $3c^4 \times 2c = 6c^{4+1} = 6c^5$ ***Note***: $2c$ *means* $2c^1$

(c) $4ab^4 \times 2a^2b^3 = 8a^{1+2}b^{4+3}$

$= 8a^3b^7$

2. **Quotient Rule states:**

   $a^m \div a^n = a^{m-n}$

Example:

Simplify

(a) $a^9 \div a^4 = a^{9-4} = a^5$

(b) $12c^4 \div 2c = 6c^{4-1} = 6c^3$

(c) $2a^3b^4 \div 5a^2b = 0.4a^{3-2}b^{4-1}$

$= 0.4a^1b^3$

3. **Power to Power Rule states:**

   $(a^m)^n = a^{m \times n}$ So, $(2^2)^3 = 2^{2\times3} = 2^6$ which simplify to give = 32.

This rule is saying you are raising the index form within the bracket to a power which is outside the bracket. With the Power-to-Power rule, we are taking the power outside the bracket and multiply it by the power of ALL the quantities within the bracket. It must be noted with the Power-to-Power rule, the constant has a power of 1.

Additionally, Power-to-Power rule can be converted to a Product rule format. Such as:

$(a^5)^2 = a^5 \times a^5 = a^{5+5} = a^{10}$. The outside index tells you how many times to multiply the insides term repeatedly.

Example:

Simplify

1. $(a^6)^2$

Working:

$a^{6\times2} = a^{12}$

OR converting to Product Rule Form:

$(a^6)^2$

$a^6 \times a^6 = a^{6+6} = a^{12}$

2. $(5a^3b^2)^3$

Working:

$$(5a^3b^2)^3 = (5^1a^3b^2)^3$$
$$= 5^{1\times3}a^{3\times3}b^{2\times3}$$
$$= 5^3a^9b^6$$
$$= 125a^9b^6$$

OR converting to Product Rule Format:

$$(5a^3b^2)^3 = 5a^3b^2 \times 5a^3b^2 \times 5a^3b^2$$
$$= 125a^{3+3+3}b^{2+2+2}$$
$$= 125a^9b^6$$

4. Zero Index

Now $5^2 \div 5^2 = 5^{2-2} = 5^0$

However, $5^2 \div 5^2$ *also means* $25 \div 25 = 1$

Additionally, $5^2 \div 5^2$ *also means* $\frac{5\times5}{5\times5} = \frac{1}{1} = 1$

Therefore, $5^0 = 1$

Also:

Now $a^2 \div a^2 = a^{2-2} = a^0$

However, $a^2 \div a^2$ *also means* $(a \times a) \div (a \times a) = 1$

Additionally, $a^2 \div a^2$ *also means* $\frac{a\times a}{a\times a} = \frac{1}{1} = 1$

Therefore, $a^0 = 1$

Hence, any quantity raised to the power of zero or raised to index zero is equal to 1.

Example:

Determine or evaluate the value of the following:

1. 19^0 2. $(14)^0$ 3. n^0 4. $\left(\frac{2}{5}\right)^0$

Working:

1. $19^0 = 1$ 2. $(14)^0 = 1$ 3. $n^0 = 1$ 4. $\left(\frac{2}{5}\right)^0 = 1$

5. Fractional (Rational) Index

$a^{\frac{m}{n}}$, *is an exmaple of a Fractional Index. That is; the power is a fraction.*

Note:

Any quantity with the fractional (rational index); the denominator is the root, and the numerator is power to which the quantity is to be raised.

Example:

(1) $4^{\frac{3}{2}}$ means the square root of the cube of 4.

(2) $8^{\frac{2}{3}}$ means the cube root of the square of 8.

Now:

$a^{\frac{m}{n}}$ *can be written in what is known as a* ***RADICAL Form***. The symbol, $\sqrt{\ }$ is called the radical symbol. *This radical form in which* $a^{\frac{m}{n}}$ *can be rewritten is*:

$\sqrt[n]{a^m}$

The denominator (root) goes outside the root symbol while the numerator (power) goes inside the root symbol raising the base 'a'.

Example:

Evaluate

1. $27^{\frac{2}{3}}$ 2. $16^{\frac{3}{2}}$

3. $(25a^6)^{\frac{3}{2}}$ 4. $(8x^9)^{\frac{2}{3}}$

Workings:

1. $27^{\frac{2}{3}} = \sqrt[3]{27^2} = 9$ 2. $16^{\frac{3}{2}} = \sqrt[2]{16^3} = 64$

3. $(25a^6)^{\frac{3}{2}} = (25^1a^6)^{\frac{3}{2}} = 25^{1\times\frac{3}{2}}a^{6\times\frac{3}{2}}$
$= 25^{\frac{3}{2}}\,a^9 = \sqrt[2]{25^3}\ a^9$
$Answer = \ 125a^9$

4. $(8x^9)^{\frac{2}{3}} = (8^1x^9)^{\frac{2}{3}} = 8^{1\times\frac{2}{3}}x^{9\times\frac{2}{3}}$
$= 8^{\frac{2}{3}}\,x^6 = \sqrt[3]{8^2}\ x^6$
$Answer = \ 4x^6$

6. Negative Index

Now $5^3 \div 5^5 = 5^{3-5} = 5^{-2}$

However, $5^3 \div 5^5\ can\ be = \frac{5\times5\times5}{5\times5\times5\times5\times5} = \frac{1}{5^2}$

Therefore, $5^{-2} = \frac{1}{5^2}$ if required a positive index

This means then that:
$a^{-n} = \frac{1}{a^n}$ if required a positive index

Note: A quantity with a negative index is the inverse (or reciprocal) of the quantity with a positive index of the same magnitude.

Example:
Rewrite the following expressions using positive index only:

1. 7^{-2}
2. x^{-5}
3. 2^{-3}

Working:
1. $7^{-2} = \frac{1}{7^2}$ simplifying to give $\frac{1}{49}$
2. $x^{-5} = \frac{1}{x^5}$
3. $2^{-3} = \frac{1}{2^3}$ simplifying to give $\frac{1}{8}$

3. Rewrite the expression a^2b^{-3}using positive index only

Working:
$a^2b^{-3} = \frac{1a^2}{b^3}$

Note: *a^2 has a positive index so it goes to the numerator.*
The b^{-3} has a negative index, so it goes to the denominator.

Activity 12a

Express each of the following in index form:

1. $2 \times 2 \times 2 \times 2$
2. $4 \times 4 \times 4 \times 4 \times 4 \times 4$
3. $y \times y \times y$
4. $a \times a \times a \times a \times a$
5. $2^4 \times 2^6$
6. $3^2 \times 3^4 \times 3^5$
7. $xy^4 \times x^3y^2$
8. $6^7 \div 6^2$
9. $\frac{4^9}{4^2}$
10. $p^7 \div p^3$
11. $a^7b^3 \div a^2b$
12. $\frac{x^4y^8}{x^3y^2}$

Activity 12b

Simplify

1. $(2^3)^2$
2. $(3^2)^3$
3. $(2a^3)^2$
4. $(3n^2b)^3$
5. $\left(\frac{2^4}{3^3}\right)^2$
6. $\left(\frac{1^4}{5^2}\right)^2$
7. $\left(\frac{y^4}{x^5}\right)^2$
8. $\left(\frac{2^4}{y^5}\right)^2$
9. $\left(\frac{b^4}{xy^5}\right)^2$
10. $\left(\frac{x^2y^3}{n^5}\right)^2$

Activity 12c

Determine the value of the following:

1. 9^0
2. $(125)^0$
3. $(y)^0$
4. 44^0
5. $\left(\frac{7}{5}\right)^0$

Activity 12d

Write down each of the following using positive index then simplify if possible

1. 5^{-3}
2. $(3)^{-2}$
3. $(7)^{-2}$
4. x^{-2}
5. $(y)^{-4}$
6. $a^{-3}b^2$

Activity 12e

Evaluate

1. $4^{\frac{1}{2}}$
2. $16^{\frac{1}{2}}$
3. $81^{\frac{1}{2}}$
4. $144^{\frac{1}{2}}$
5. $1^{\frac{1}{2}}$
6. $1^{\frac{1}{3}}$
7. $27^{\frac{1}{3}}$
8. $8^{\frac{1}{3}}$
9. $125^{\frac{1}{3}}$
10. $8^{\frac{4}{3}}$
11. $64^{\frac{2}{3}}$
12. $(125m^3)^{\frac{2}{3}}$
13. $(36y^4)^{\frac{3}{2}}$
14. $(27x^9)^{\frac{2}{3}}$
15. $(16n^6)^{\frac{3}{2}}$

Transposition – Subject of the Formula

(Where the Subject Appear Once)

A formula is a general equation which shows the connection between two or more related quantities.

For example: $A = \pi r^2$ *is the formula for the area of a circle.* In that formula A which represents the area, is the subject of the formula.

To make a quantity the subject means, every other quantity must be equal to the required quantity. That is, **changing the subject of the formula or transposing the formula** refers to the process whereby the formula is rearranged so that the required symbol becomes the subject.

Note: In transposition of formulae, ALL the rules are obeyed as used in solving equations.

It should also be noted that it is best when making a symbol the subject of the formula that the following be observed:

- **(i)** **Let the subject be on the L.H.S. of the formula**
- **(ii)** **The subject must be in the numerator**
- **(iii)** **The power of the subject must be 1**
- **(iv)** **The subject must be positive**
- **(v)** **The coefficient of the subject must be 1**

Examples:

1. *If* $a = b + c$, *make* c *the subject.*

Working:

$a = b + c$ *is the same as* $b + c = a$

So, $b + c = a$

$+c = a - b$

$c = a - b$

2. *If* $a = b - c$, *transpose for* c

Working:

$b - c = a$

So, $-c = a - b$; **Note:** c MUST NOT be $-ve$

So, $c = \frac{a-b}{-1}$

Hence, $c = -a + b$

3. *If* $u + at = v$, *make* $\mathbf{t}$ *the subject*

Working:

$u + at = v$

$+at = v - u$

$t = \frac{v-u}{a}$

It is best to get rid of an operation *first that is not connected to the subject by multiplication.*

4. *If* $ax = b$, *transpose for* x

Working:

$ax = b$ **Note**: *ax means* $a \times x$

$x = \frac{b}{a}$

5. $If\ abc = m, transpose\ for\ b$

Working:

$abc = m$ $\textbf{Note}: abc\ means\ a \times b \times c$

$b = \frac{m}{ac}$

6. $Transpose\ I = \frac{v}{R}; for\ V$

Working:

$\frac{V}{R} = I$

$V\ is\ at\ the\ numerator, so\ get\ rid\ of\ R$

$by\ multiplying\ by\ R\ on\ the\ R.H.S.$

$V = I(R)$

$V = IR$

7. $If\ x^2 = y,\ transpose\ for\ x$

Working:

$x^2 = y$ $\textbf{Note}: the\ power\ of\ x\ is\ not\ 1.$

$x = \sqrt{y}$ $The\ inverse\ of\ squaring\ is\ square\ root$

8. $If\ \sqrt{x} = y,\ transpose\ for\ x$

Working: $\textbf{Note}: the\ power\ of\ x\ is\ not\ 1$

$\sqrt{x} = y$ $The\ \sqrt{\ }\ symbol\ mean\ power\ \frac{1}{2}$

$x = (y)^2$ $The\ inverse\ of\ square\ root\ is\ square.$

9. $\sqrt{\frac{A}{m}} = v;\ make\ A\ the\ subject$

Working:

$\sqrt{\frac{A}{m}} = v$ $\textbf{Note}: 1st\ we\ remove\ the\ square\ root\ symbol$

$\frac{A}{m} = (v)^2$ $to\ get\ closer\ to\ A.\ Also\ because\ the\ square\ root\ symbol\ covers\ the\ entire\ \frac{A}{m}$

$A = (v)^2 m$ $\textbf{Note}: 2nd\ we\ remove\ m\ by\ multiplying\ because\ A\ is\ at$

$at\ the\ numerator.$

10. $Given\ \frac{a}{b+c} = x,\ make\ b\ the\ subject$

Working:

$\frac{a}{b+c} = x$ $\textbf{Note}: x\ means\ \frac{x}{1}$

$\textbf{Note}: The\ subject\ b\ is\ at\ the\ denominator.$
$The\ subject\ MUST\ be\ at\ the\ numerator.$

So, we invert both sides [L.H.S. & R.H.S.] to give:

$\frac{b+c}{a} = \frac{1}{x}$

$Now, b\ is\ at\ the\ numeator, so\ we\ remove\ the\ denominator$

first to give:

$b + c = \left(\frac{1}{x}\right) \times a$

$b + c = \left(\frac{a}{x}\right)$

11. If $u^2 + 2ax = v^2;\ for\ x$

Working:

$+2ax = v^2 - u^2$ $Getting\ rid\ of\ u^2\ first.$

$x = \frac{v^2 - u^2}{2a}$ $[see\ reason\ given\ by\ example\ 3].$

12. $If\ y^2 - a = b^2;\ for\ y$

Working:

$y^2 = b^2 + a$

$y = \sqrt{(b^2 + a)}$

$b = \left(\frac{a}{x}\right) - c$

Activity 13

1. $Given\ the\ formula\ mx + c = y; make\ m\ the\ subject.$
2. $Given\ the\ formula\ m = \frac{4th^2}{x}; rewrite\ the\ formula\ to\ make\ t\ the\ subject.$
3. $For\ the\ formula\ A = 2x^2 + 4yh; make\ h\ the\ subject\ of\ the\ formula.$
4. $Transpose\ the\ formula\ T = \frac{2x}{y}\ to\ make\ y\ the\ subject.$
5. $A\ formula\ for\ velocity\ is\ v = \sqrt{2ax + u^2}; make\ \boldsymbol{a}\ the\ subject.$
6. $Given\ ab = c^2d; make\ c\ the\ subject\ of\ the\ formula.$
7. $Given\ x(a + b) = c; rearrange\ the\ formula\ to\ make\ b\ the\ subject.$
8. $For\ the\ formula\ A = C + Crt, make\ t\ the\ subject\ of\ the\ formula.$
9. $A\ formula\ for\ the\ area\ of\ a\ trapezium\ is:$
$$A = \frac{h(a + b)}{2}$$
$Rearrange\ the\ formula\ to\ make\ b\ the\ subject.$

10. $A\ formula\ for\ acceleration\ is:$
$$a = \frac{v^2 - u^2}{2x}$$
$Rearrange\ the\ formula\ above\ to\ make\ u\ the\ subject.$

Factorization by Grouping

This method normally gives four (4) algebraic terms to factorize. Firstly, we group the algebraic terms in pairs so that each pair of terms has a common factor. We then factor out the common factor for each pair of terms. After a common factor for the pair of factorized terms is found, factor it out and then the process of factorization is completed. Here are a few illustrated examples below.

Example:

Factorize each of the following algebraic expressions

1. $ax + ay + bx + by$

Now

$Step\ 1$ $(ax + ay) + (bx + by),$ Grouping in pairs

$Step\ 2$ $a(x + y) + b(x + y)$ Factoring out the common factor in each pair.

Answer $= (x + y)(a + b)$ Now factoring out $(x + y)$, which is common to both terms, seen in Step 2. $The\ a\ with\ + b\ is\ the\ other\ factor\ we\ then\ put\ in\ bracket.$

2. $2px - 6py + bx - 3by$

   Now

   $Step\ 1 \quad (2px - 6py) + (bx - 3by),$ Grouping in pairs

   $Step\ 2 \quad 2p(1x - 3y) + 1b(1x - 3y)$ Factoring out the common factor in each pair.

   Answer $= (1x - 3y)(2p + 1b)$ Now factoring out $(1x - 3y)$, which is common to both terms seen in Step 2.

3. $5px - 5py - 3qx + 3qy$

   Now

   $Step\ 1 \quad (5px - 5py) - (3qx - 3qy),$

   Grouping in pairs. However, whenever you place the bracket around a pair of terms; if a NEGATIVE sign is outside the bracket, it will change the original given sign inside the bracket. Note, we had from the original expression for the last two terms: $-3qx + 3qy, when\ we\ put\ the\ bracket\ such\ as - (3qx - 3qy); it\ can\ be\ seen\ we\ had\ a + 3qy\ before, now\ it\ is - 3qy.$

   It is because you are multiplying the bracket out by -1, which when expand will give back $-3qx + 3qy$.

   $Step\ 2 \quad 5p(x - y) - 3q(x - y)$ ← Factoring out the common factor in each pair

   Answer $= (x - y)(5p - 3q)$ Now factoring out $(x - y)$, which is common to both terms seen in Step 2.

4. $ax + bx - ay - by$

   Now

   $= (ax + bx) - (ay + by)$

   $= x(a + b) - y(a + b)$

   $Answer = (a + b)(x - y)$

5. $ab(5x - 2) + 2(5x - 2)$

   $This\ is\ partcially\ factorized\ and\ common\ factor\ is\ (5x - 2)$

   $Answer = (5x - 2)(ab + 2)$

Activity 14

$Factorize\ each\ of\ the\ following\ algebraic\ express$ions

1. $mx + my + nx + ny$
2. $ab + cb + ad + cd$
3. $ax + 5bx + ay + 5by$
4. $2ax - 8ay + bx - 4by$
5. $ap - 4np + aq - 4nq$
6. $4xr + 4xt - 5yr - 5yt$
7. $2ax - 2ay - 3bx + 3by$
8. $3rx - 3ry - 5bx + 5by$

Expanding the Product of Two Binomial Expressions

Consider $(x + a)(x + b), is\ a\ Binomial\ expression\ consisting\ of\ two\ terms.$

Their product will be according to the distributive law where:

$(x + a)(x + b) = x(x + b) + a(x + b) = x^2 + bx + ax + ab.$

We took the first binomial term $x + a\ and\ use\ each\ single\ term$ to multiply by the second binomial term $(x + b)$.

The algebraic expression $x^2 + bx + ax + ab, is\ a\ \boldsymbol{QUADRATIC\ EXPRESSION}.$ $This\ is\ because\ the\ highest\ power\ of\ the\ x\ in\ the\ expression\ is\ 2.$

Note: The product of two binomial expressions is NOT always a quadratic expression. For example: $(a+b)(c+d) = a(c+d) + b(c+d) = ac + ad + bc + bd$, which is NOT Quadratic.

$x^2 + bx + ax + ab$ *can also be written in the general form* $x^2 + (a+b)x + ab$.

Note: $(x+y)^2$ *means* $(x+y)(x+y) \Rightarrow x(x+y) + y(x+y) = x^2 + xy + xy + y^2$
$= x^2 + 2xy + y^2$

Example:

Expand and simplify each of the following expressions:

1. $(x+4)(x+5)$

   Working:

   $x(x+5) + 4(x+5)$

   $x^2 + 5x + 4x + 20$

   $x^2 + 9x + 20$

2. $(x+2)(x-3)$

   Working:

   $x(x-3) + 2(x-3)$

   $x^2 - 3x + 2x - 6$

   $x^2 - 1x - 6$

3. $(x-5)(x+7)$

   Working:

   $x(x+7) - 5(x+7)$

   $x^2 + 7x - 5x - 35$

   $x^2 + 2x - 35$

4. $(x-5)(x-3)$

   Working:

   $x(x-3) - 5(x-3)$

   $x^2 - 3x - 5x + 15$

   $x^2 - 8x + 15$

Activity 15

Expand and simplify the following

1. $(x+3)(x+2)$
2. $(x+6)(x+1)$
3. $(x-7)(x+1)$
4. $(x-2)(x+8)$
5. $(x-5)(x-2)$
6. $(x-4)(x-3)$
7. $(2x+1)(x+3)$
8. $(x+3)(2x-4)$
9. $(x+5)^2$
10. $(x-3)^2$
11. $(2x+1)^2$
12. $(2-2x)^2$

Factorizing Quadratic Expressions

Note: All expression whereby the highest degree in that statement is 2 are called Quadratic. Hence, expressions such as:

(1) $ax^2 \pm bx$

(2) $ax^2 - c$ (difference of two squares)

(3) $ax^2 \pm bx \pm c$

are All Quadratic expressions, because x^2 *is the highest degree (power)*, in those expressions.

Now, depending on the type of expression; when required to factorize, will determine how many brackets will be used. Type (1) $ax^2 \pm bx$, will require one bracket. While type (2) $ax^2 - c$ and type (3) $ax^2 \pm bx \pm c$ will require two brackets. So, if the quadratic expressions have a x^2 *term and c (constant), then when factorize, there will be two brackets involved.*

Example:

Factorize completely

1. $12x^2 + 8x$

   Working:

   Using factorizing by H.C.F.

   $H.C.F. = 4x$

   $\cancel{4x}.3x + \cancel{4x}.2$

   $= 4x(3x + 2)$

2. $6x^2 - 3x$

   Working:

   Using factorizing by H.C.F.

   $H.C.F. = 3x$

   $\cancel{3x}.2x - \cancel{3x}.1$

   $= 3x(2x - 1)$

3. $-5x + 20x^2$

   Working:

   Using factorizing by H.C.F.

   $H.C.F. = 5x$

   $\cancel{5x}.-1 + \cancel{5x}.4x$

   $= 5x(-1 + 4x)$

Activity 16a

Factorize completely

1. $12x + 9x^2$
2. $6x^2 - 8x$
3. $30x^2 + 18x$
4. $8x^2 + 20x$
5. $15x - 20x^2$
6. $7x + 14x^2$

Example:

Factorize the following quadratic expressions:

1. $x^2 + 5x + 2x + 10$

   Using the concept of factorizing by grouping

   $(x^2 + 5x) + (2x + 10)$

   $\Rightarrow x(x + 5) + 2(x + 5)$

   $= (x + 5)(x + 2)$

2. $x^2 + 4x - 3x - 12$

   Using the concept of factorizing by grouping

   $(x^2 + 4x) - (3x + 12)$

   $\Rightarrow x(x + 4) - 3(x + 4)$

   $= (x + 4)(x - 3)$

3. $6x^2 + 10x + 3x + 5$

   Using the concept of factorizing by grouping

   . $(6x^2 + 10x) + (3x + 5)$

   $\Rightarrow 2x(3x + 5) + 1(3x + 5)$

   $= (3x + 5)(2x + 1)$

4. $3 + 5x - 6x - 10x^2$

   Using the concept of factorizing by grouping

   $(3 + 5x) - (6x + 10x^2)$

   $\Rightarrow 1(3 + 5x) - 2x(3 + 5x)$

   $= (3 + 5x)(1 - 2x)$

Note: From the above expressions 1 – 4

We use factorization by grouping because we had four terms involved.
Also you could see each grouping pair had a common factor involved.

Activity 16b

1. $x^2 + 4x + 3x + 12$
2. $x^2 + 4x + 5x + 20$
3. $x^2 - 7x - 6x + 42$
4. $x^2 - 6x - 5x + 30$
5. $x^2 - x + 4x - 4$
6. $x^2 - 9x + 8x - 72$
7. $6 + 3x - 2x - x^2$
8. $3 - 4x + 6x - 8x^2$
9. $12x^2 + 9x + 4x + 3$
10. $10 - 14x - 15x + 21x^2$

Example:

Factorizing the quadratic type $ax^2 \pm bx \pm c \ when \ a = 1$

Factorize completely

1. $x^2 + 8x + 12$

Working:

$1x^2 + 8x + 12$

$\mathbf{1 \times 12 = 12}$

Find two integers when multiply gives $\mathbf{12}$ *and when Add gives the middle term coefficient of x. These integers are* 6 & 2. *That is,* $\mathbf{6 \times 2 = 12}$ & $\mathbf{6 + 2 = 8}$

Now expressing the expression into four terms by splitting $8x$ *into* $\mathbf{+6x + 2x}$; *see below*

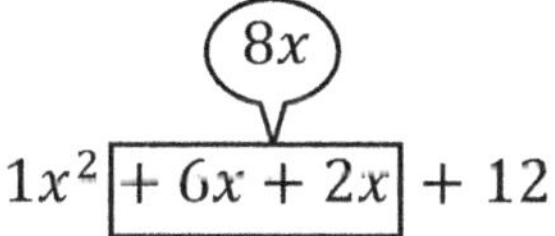

Using the concept of factorizing by grouping

$$(1x^2 + 6x) + (2x + 12)$$
$$1x(x + 6) + 2(x + 6)$$
$$= (x + 6)(1x + 2)$$

2. $x^2 + 5x - 14$

Working:

$1x^2 + 5x - 14$

$\mathbf{1 \times -14 = -14}$

Find two two integers when multiply gives $\mathbf{-14}$ *and when Add gives the middle term coefficient of x. These are* -2 & 7. *That is,* $\mathbf{-2 \times 7 = -14}$ & $\mathbf{-2 + 7 = 5}$

Now expressing the expression into four terms by splitting $5x$ *into* $\mathbf{-2x + 7x}$; *see below*

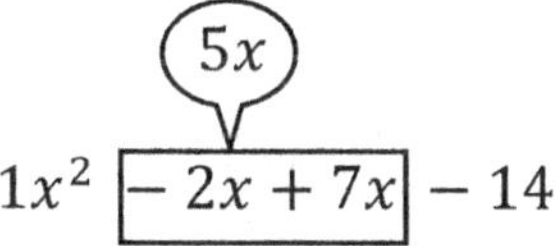

Using the concept of factorizing by grouping

$$(1x^2 - 2x) + (7x - 14)$$
$$1x(x - 2) + 7(x - 2)$$
$$= (x - 2)(1x + 7)$$

3. $x^2 - 8x + 15$

Working:

$1x^2 - 8x + 15$

$1 \times 15 = 15$

Find two integers when multiply gives $\mathbf{15}$ *and when Add gives the middle term coefficient of x. These integers are* -3 & -5. *That is,* $\mathbf{-3 \times -5 = +15}$ & $-3 + -5 = -8$. *But* $-3 + -5$ *is the same as* $\mathbf{-3 - 5 = -8}$.

Now expressing the expression into four terms by splitting $-8x$ *into* $\mathbf{-3x - 5x}$; *see below*

$$1x^2 \boxed{-3x - 5x} + 15$$
($-3x - 5x = -8x$)

Using the concept of factorizing by grouping

$(1x^2 - 3x) - (5x - 15)$

$\Rightarrow 1x(x-3) - 5(x-3)$

$= (x-3)(1x-5)$

Example:

Factorizing the quadratic type $ax^2 \pm bx \pm c$ $when\ 1 < a < 1; but\ a \neq 0$

Note: $when\ a = -1; the\ concept\ is\ basically\ the\ same\ as\ when\ a = 1$. It is just to find out what integers is possible to get the middle coefficient of x.

Factorize completely

1. $12x^2 + 17x + 6$

Working:

$12x^2 + 17x + 6$

$12 \times 6 = 72$

The aim is to make the coefficient of x^2 be 1. So we take 72 and replace it with the constant + 6 and so the coefficient of x^2 is now 1; after we had multiplied by 12 by 6. So we now have:

$1x^2 + 17x + 72$

So:

$\Rightarrow 1x^2 + 8x + 9x + 72$

$\Rightarrow 1x(x+8) + 9(x+8)$

$\Rightarrow (x+8)(1x+9)$

However, it must be noted that because we had changed the original equation by multiplying by 12 then we MUST reverse the proces by dividing by 12 into the two constant + 8 and + 9.

$\Rightarrow \left(x + \frac{8}{12}\right)\left(x + \frac{9}{12}\right) = \left(x + \frac{2}{3}\right)\left(x + \frac{3}{4}\right)$

Taking each denominator and put as the coefficient of x

Answer $= (3x+2)(4x+3)$

Factorize completely

2. $2x^2 + 5x - 12$

Working:

$2x^2 + 5x - 12$

$2 \times -12 = -24$

The aim is to make the coefficient of x^2 be 1. So we take − 24 and replace it with the constant − 12 and so the coefficient of x^2 is now 1, after we had multiplied by 2 by − 12. So we now have:

$1x^2 + 5x - 24$

$1x^2 + 8x - 3x - 24$

$\Rightarrow (1x^2 + 8x) - (3x + 24)$

$1x(x+8) - 3(x+8)$

$\Rightarrow (x+8)(1x-3)$

However, it must be noted that because we had changed the original equation by multiplying by 2 then we MUST reverse the proces by dividing by 2 into the two constant + 8 and − 3.

$\Rightarrow \left(x + \frac{8}{2}\right)\left(x - \frac{3}{2}\right) = \left(x + \frac{4}{1}\right)\left(x - \frac{3}{2}\right)$

Taking each denominator and put as the coefficient of x

$Answer = (1x+4)(2x-3)$

Activity 16c

Factorize completely

1. $15x^2 + 26x + 8$
2. $16x^2 + 8x + 1$
3. $12x^2 - x - 6$
4. $2x^2 - 7x + 6$
5. $3x^2 - 11x - 4$
6. $6x^2 + 7x - 3$
7. $3x^2 - 5x + 2$
8. $8x^2 + 10x + 3$
9. $15x^2 - 4x - 3$
10. $6x^2 + 7x - 20$

Factorizing The Difference of Two Square:

As indicated:

difference means to subtract, so the statement is suggesting to factorize two square terms that are subtracting. For examples: $(1)x^2 - 16$ $(2)4x^2 - 9$.

Now:

$(1)\ x^2 - 16$ *or* $1x^2 - 16$ *shows All components of each term are squared.*

That is $1 \times 1 \times x \times x - 4 \times 4 = 1x^2 - 16$ showing All components are squared.

$(2)\ 4x^2 - 9$ *shows All components of each term are squared.*

That is $2 \times 2 \times x \times x - 3 \times 3 = 4x^2 - 9$ showing All components are squared.

Example:

Factorize completely

1. $x^2 - 16$

Working:

1st on recognizing x^2 *and 16 are both square terms and they are subtrating.*

We do the following:

(+4)

$$x \,.\, x - 4 \,.\, 4 \;=\; x \,.\, (x - 4) \,.\, 4$$

$(x + 4)$

Answer $= (x - 4)(x + 4)$

2. $9x^2 - 25$

Working:

1st on recognizing $9x^2$ *and 25 are both square terms and they are subtractig.*

We do the following:

$$3x \,.\, 3x - 5 \,.\, 5 = 3x \,.\, (3x - 5) \,.\, 5$$

$(3x + 5)$

Answer $= (3x - 5)(3x + 5)$

Activity 16d

Factorize the following:

1. $x^2 - 144$
2. $x^2 - 1$
3. $x^2 - 100$
4. $x^2 - 49$
5. $x^2 - 81$
6. $9x^2 - 1$
7. $25x^2 - 1$
8. $36x^2 - 1$
9. $4x^2 - 9$
10. $9x^2 - 36$

Algebraic Fractions

Addition and Subtraction – (Binomial Numerator)

Example:

Simplify

1. $\frac{x+5}{3}+\frac{x+4}{2}$

Working:

L.C.D. of 3 & 2 is 6

$$\frac{2(x+5)}{2(3)}+\frac{3(x+4)}{3(2)}$$

$$=\frac{2(x+5)+3(x+4)}{6}$$

$$=\frac{2x+10+3x+12}{6}$$ *Grouping Like Terms*

$$\Rightarrow\frac{2x+3x+10+12}{6}=\frac{5x+22}{6}$$

2. $\frac{x+1}{4}-\frac{x-7}{12}$

Working:

L.C.D. of 4 and 12 is 12

$$\frac{3(x+1)}{3(4)}-\frac{1(x-7)}{1(12)}$$

$$=\frac{3(x+1)-1(x-7)}{12}$$

$$=\frac{3x+3-1x+7}{12}$$ *Grouping Like Terms*

$$\Rightarrow\frac{3x-1x+3+7}{12}=\frac{2x+10}{12}$$

Simplifying because all terms are multiples of 2 by dividing throughtout by 2.

$$=\frac{x+5}{6}$$

Addition and Subtraction – (Binomial Denominator/or Variable)

Note:

$3\times 5 \text{ or } 3(5)\equiv 5\times 3 \text{ or } 5(3)\equiv 15$

$(a+b)(x+y)\equiv(x+y)(a+b)$

$a\times b \text{ or } a(b)\equiv b\times a \text{ or } b(a)\equiv ab$

$(a+b)(x)\equiv(x)(a+b)$

Example:

Simplify

1. $\frac{3}{x+2}-\frac{5}{x-7}$

Working:

If you have at least a binomial denominator.
The L.C.D. is a product of the two denominator.
So the L.C.D. is $(x+2)(x-7)$

$$=\frac{3(x-7)}{(x+2)(x-7)}-\frac{5(x+2)}{(x+2)(x-7)}$$

2. $\frac{9}{x}+\frac{5}{x+4}$

Working:

If you have at least a binomial denominator.
The L.C.D. is a product of the two denominator.
So the L.C.D. is $(x)(x+4)$

$$=\frac{9(x+4)}{(x)(x+4)}+\frac{5(x)}{(x)(x+4)}$$

$$= \frac{3(x-7)-5(x+2)}{(x+2)(x-7)}$$

$$= \frac{3x-21-5x-10}{(x+2)(x-7)}$$ *Grouping Like Terms*

$$= \frac{3x-5x-21-10}{(x+2)(x-7)}$$

$$= \frac{-2x-31}{(x+2)(x-7)}$$

$$= \frac{9(x+4)+5(x)}{(x)(x+4)}$$

$$= \frac{9x+36+5x}{(x)(x+4)}$$ *Grouping Like Terms*

$$= \frac{9x+5x+36}{(x)(x+4)}$$

$$= \frac{14x+36}{(x)(x+4)}$$

3. $\frac{3}{4} + \frac{7}{x}$

Working: $L.C.D. \text{ is } 4x$

$$\frac{3}{4} + \frac{7}{x}$$

$$= \frac{3(x)}{4(x)} + \frac{4(7)}{4(x)} = \frac{3(x)+4(7)}{4x}$$

$$= \frac{3x+28}{4x}$$

4. $\frac{3}{x} - \frac{5}{7x}$

Working: $L.C.D. \text{ is } 7x$

$$\frac{3}{x} - \frac{5}{7x}$$

$$= \frac{7(3)}{7(x)} - \frac{1(5)}{1(7x)} = \frac{7(3)-1(5)}{7x}$$

$$= \frac{21-5}{7x} = \frac{16}{7x}$$

5. $\frac{5}{9x} + \frac{2}{3x^2}$

Working: $L.C.D. \text{ is } 9x^2$

$$\frac{5}{9x} + \frac{2}{3x^2}$$

$$= \frac{5(x)}{9x(x)} + \frac{3(2)}{3(3x^2)} = \frac{5(x)+3(2)}{9x^2}$$

$$Answer = \frac{5x+6}{9x^2}$$

Simplify

6. $\frac{3}{x-5} + \frac{2}{x^2-2x-15}$

Working:

We will factorize $x^2 - 2x - 15$ *to give* $(x-5)(x+3)$
Hence, rewriting the above expression gives:

$$\frac{3}{x-5} + \frac{2}{(x-5)(x+3)}$$

Now the L.C.D is $(x-5)(x+3)$

Note: *If the L.C.D. for two numbers 3 and 5 is 15. We can say 15 is the same as* $(3)(5) = (5)(3) = 15$. *So the L.C.D. of* $(x-5)$ & $(x-5)(x+3)$ *is* $(x-5)(x+3)$

$$\frac{3(x+3)}{(x-5)(x+3)}+\frac{2}{(x-5)(x+3)}=\frac{3(x+3)+2}{(x-5)(x+3)}$$

$$=\frac{3x+9+2}{(x-5)(x+3)}$$

$$=\frac{3x+11}{(x-5)(x+3)}$$

Activity 17

Simplify the following

1. $\frac{4}{7x}+\frac{3}{5x}$
2. $\frac{3}{5x}-\frac{2}{9x}$
3. $\frac{7}{12x}+\frac{5}{8}$
4. $\frac{3}{2x}+\frac{2}{5}$
5. $\frac{9}{5xy}-\frac{8}{15x}$
6. $\frac{2}{3x}-\frac{4}{5x^2}$
7. $\frac{6}{7y^2}+\frac{5}{14y}$
8. $\frac{4}{3n}-\frac{5}{2mn}$
9. $\frac{x+3}{4}+\frac{x+2}{3}$
10. $\frac{x+5}{6}-\frac{x+4}{12}$
11. $\frac{x-2}{3}-\frac{x-3}{4}$
12. $\frac{x+9}{8}-\frac{x}{4}$
13. $\frac{7}{x}+\frac{3}{x+5}$
14. $\frac{1}{x+3}+\frac{2}{x}$
15. $\frac{4}{x+2}+\frac{9}{x-5}$
16. $\frac{8}{x-2}-\frac{4}{x}$
17. $\frac{7}{x-5}-\frac{3}{x-4}$
18. $\frac{5}{x+1}-\frac{6}{x-3}$
19. $\frac{3}{x-1}-\frac{8}{x+5}$
20. $\frac{12}{x}-\frac{7}{x+5}$
21. $\frac{7}{x+1}+\frac{9}{(x+1)(x+2)}$
22. $\frac{4}{(x-7)(x-2)}-\frac{3}{(x-2)}$
23. $\frac{12}{x-6}-\frac{5}{x^2-3x-18}$
24. $\frac{1}{x-7}+\frac{4}{x^2-49}$

Multiplication and Division of Algebraic Fractions

Multiplication

When multiplying algebraic fractions, firstly we multiply the numerators together then we multiply the denominators together. After finding the products of the numerator and the products of the denominators we then reduce the fraction by cancelling common factors to both the numerator and denominator.

Example:

Simplify

1. $\frac{m^4}{b^2c^3} \times \frac{b^3c^2}{m}$

2. $\frac{3a^4b}{7bc} \times \frac{14c^3}{a^2b^2}$

Working:

$$\Rightarrow \frac{m^4}{b^2c^3} \times \frac{b^3c^2}{m} = \frac{m \times m \times m \times m \times b \times b \times b \times c \times c}{b \times b \times c \times c \times c \times m}$$

$$= \frac{m^3b}{c}$$

Working:

$$\Rightarrow \frac{3a^4b}{7bc} \times \frac{14c^3}{a^2b^2} = \frac{3 \times 14 \times a \times a \times a \times a \times b \times c \times c \times c}{7 \times b \times b \times b \times a \times a \times c}$$

$$= \frac{6a^2c^2}{b^2}$$

Division

Under division of algebraic fractions, we invert the fraction which is the divisor after we had changed the division sign to multiplication. Afterwards, we follow the same procedure as did with multiplication of algebraic fractions.

Example:

Simplify

1. $\frac{a^3b}{cd} \div \frac{a^5}{c^2r^2}$

2. $\frac{2ab}{9rx} \div \frac{4a^3}{9r^2}$

Working:

$\frac{a^3b}{cd} \div \frac{a^5}{c^2r^2}$

$$\Rightarrow \frac{a^3b}{cd} \times \frac{c^2r^2}{a^5} = \frac{a \times a \times a \times b \times c \times c \times r \times r}{c \times d \times a \times a \times a \times a \times a}$$

$$= \frac{bcr^2}{a^2d}$$

Working:

$\frac{2ab}{9rx} \div \frac{4a^3}{9r^2}$

$$\Rightarrow \frac{2ab}{9rx} \times \frac{9r^2}{4a^3} = \frac{2 \times 9 \times a \times b \times r \times r}{9 \times 4 \times r \times x \times a \times a \times a}$$

$$= \frac{br}{2a^2x}$$

Activity 18a

Simplify the following:

1. $\frac{3m^2}{4x} \times \frac{2x^2}{6m}$
2. $\frac{a^2x^3}{b^2y} \times \frac{b^3}{a^2x}$
3. $\frac{a^2}{b^3c^2} \times \frac{b^2c^3}{a^4}$
4. $\frac{p^2q^2}{r^3} \times \frac{r^2p}{q^3}$
5. $\frac{a^3x^2}{b^2y^2} \div \frac{a^2c}{by^2}$
6. $\frac{pq^2}{m^2n} \div \frac{pq}{m^3n^2}$
7. $\frac{5a^2b}{6mn^2} \div \frac{5ab^3}{12m^2b}$
8. $\frac{5ab}{r^2} \times \frac{mn}{3ab} \times \frac{6mr}{n^2}$

Example: Factorize and simplify $\frac{a^2-25}{2a+10}$

Working: We will factorize the numerator and denominator then cancel the common terms

$$\frac{a^2-25}{2a+10} = \frac{(a-5)\cancel{(a+5)}}{2\cancel{(a+5)}} = \frac{(a-5)}{2}$$

Activity 18b

Factorize and simplify

1. $\frac{a^2-4}{5a+10}$
2. $\frac{x^2-9}{5x-15}$
3. $\frac{7x+28}{x^2-16}$
4. $\frac{3y+9}{y^2-9}$
5. $\frac{3x-6}{x^2-4}$
6. $\frac{3x-6}{x^2+3x-10}$

Solving Exponential Equations

An exponential equation is an equation whereby the variable is in the exponent position.

Note:

$If\ a^x = a^y,$

$then\ x = y; since\ the\ bases\ are\ equal, then\ the\ exponents\ are\ also\ equal.$

Note:

We use this fact to solve exponential equations (where the unknown quantity is in the index position) and their bases are equal or their bases can become equalized.

Example:

1. Solve

$5^x = 125$

Working:

The variable x in the index. We see a base of 5, on the L.H.S. Now making the 125 equalized to the base 5 and when raises to a power gives back 125. This is 5^3

$\cancel{5}^x = \cancel{5}^3$

Both bases are equal so:

$x = 3$

2. Solve

$16^x = 64$

Working:

In the above exponential equation given; there is nothing that we can do with 64 on the R.H.S. using the base 16 seen on the L.H.S. with an index to give 64. This means we have to find a common base that when raised to an index gives 16 and 64. That base is 4.

$4^{2(x)} = 4^3$

$\cancel{4}^{2x} = \cancel{4}^3; \quad \Rightarrow 2x = 3$

$x = \frac{3}{2}\ or\ x = 1.5$

3. Solve

$3^{5x} = 9^{x-6}$

Working: base 9 on R.H.S. can become a base 3 like on the L.H.S. and raised to power 2 to give back 9

$3^{5x} = 3^{2(x-6)}$

$3^{5x} = 3^{2x-12}$

$5x = 2x - 12$

Solving like equations

$5x - 2x = -12$

$3x = -12$

$x = \frac{-12}{3}$

$x = -4$

4. Solve

$(2^x)(4^{2x+1}) = 64$

Working: changing the base 4 and 64 to base 2 & raising them to a power to give back 4 and 64.

$(2^x)\left(2^{2(2x+1)}\right) = 2^6$

$(2^x)(2^{4x+2}) = 2^6$

Applying the Product Rule: (keep the base and Add Power) on the L.H.S.

$2^{x+4x+2} = 2^6$

$2^{5x+2} = 2^6$

$5x + 2 = 6$

$5x = 6 - 2$

$5x = 4 \quad \Rightarrow x = \frac{4}{5} \quad or\ x = 0.8$

5. Solve $5^{2x} = \frac{1}{625}$

Working:

$5^{2x} = \frac{1}{5^4}$

The base should be at the numerator.
So, using the Laws of Indices with negative exponents. We now have:

$5^{2x} = 5^{-4}$

$2x = -4$

$x = \frac{-4}{2} \quad \Rightarrow x = -2$

Activity 19

Solve each of the following

1. $2^x = 2^6$
2. $5^{x+3} = 125$
3. $2^{4x} = 64$
4. $3^{2x} = 243$
5. $2^{2x} = \frac{1}{8}$
6. $5^{3x-1} = 25^{x-4}$
7. $5^{3x} = \frac{1}{125}$
7. $(5^{2x-1})(5^{x-7}) = 625$

Solving Quadratic Equations

There are three algebraic methods of solving quadratic equations. These methods are:
(1) Factorization (2) Completing the Square (3) Using the Quadratic Formula.

Solving Quadratic Equations by Factorization:

In order to solve any quadratic equations using this method we MUST make sure the following are observed and done.

(1) Equate the equation to zero (0).
(2) Factorize keeping the products of the two factors to equal zero.
(3) Take each factor, equate them to zero and then solve each equation.

Note:
If we have a quadratic equation whereby $(x + a)(x + b) = 0; then$ $either\ (x + a) = 0\ or\ (x + b) = 0\ which\ means\ (x + a) = (x + b) = 0.$

Example:

1. Solve the quadratic equation
$x^2 + 5x + 6 = 0$

Working:
From the above equation, rule (1) is observed.
So, we follow through with rule (2) & (3).

Factorizing gives:
$(x + 2)(x + 3) = 0$
Now:
$(x + 2) = 0 \quad or \quad (x + 3) = 0$
$x + 2 = 0 \quad or \quad x + 3 = 0$
$x = 0 - 2 \quad or \quad x = 0 - 3$
$x = -2 \quad or \quad x = -3$
$The\ solution\ is\ x = -2\ or\ x = -3$

2. Solve the quadratic equation
$2x^2 - x - 15 = 0$

Working:
From the above equation, rule (1) is observed.
So, we follow through with rule (2) & (3).

Factorizing gives:
$(2x + 5)(x - 3) = 0$
Now:
$(2x + 5) = 0 \quad or \quad (x - 3) = 0$
$2x + 5 = 0 \quad or \quad x - 3 = 0$
$2x = 0 - 5 \quad or \quad x = 0 + 3$
$2x = -5 \quad or \quad x = 3$
$x = \frac{-5}{2} = -2.5 \quad or \quad x = 3$
$The\ solution\ is\ x = -2.5\ or\ x = 3$

3. Solve $x^2 - 6x = 0$

Working:
$x(x - 6) = 0$
$So:$

$x = 0 \quad or \quad (x - 6) = 0$

4. Solve $2a^2 - 32 = 0$

Working: Factor out 2 gives:
$2(a^2 - 16) = 0$

Within the bracket, $a^2 - 16\ is\ recognized\ as\ the$

Therefore:

$x = 0 \quad or \quad x - 6 = 0$

$x = 0 \quad or \quad x = 0 + 6$

$x = 0 \quad or \quad x = 6$

$Solution\ x = 0\ or\ x = 6$

5. Solve
   $x^2 - 8x + 16 = 1$

Working:

Note: The Quadratic equation on the L.H.S. equal to 1.
However, the equation MUST equal zero.
So, we have to equate the equation to 0 by removing the 1
to the L.H.S. then simplify the equation before factorizing & solve.

So:

$x^2 - 8x + 16 - 1 = 0$

$x^2 - 8x + 15 = 0$

Now: Factorizing the above

$(x - 3)(x - 5) = 0$

Hence:

$(x - 3) = 0 \quad or \quad (x - 5) = 0$

$x - 3 = 0 \quad or \quad x - 5 = 0$

$x = 0 + 3 \quad or \quad x = 0 + 5$

$x = 3 \quad or \quad x = 5$

$Solution\ is\ x = 3\ or\ x = 5$

Difference of Two Square.

So: $2[a\,.\,a - 4\,.\,4] = 0$

$\Rightarrow 2(a - 4)(a + 4) = 0$

$\Rightarrow (a - 4)(a + 4) = \frac{0}{2}$

$\Rightarrow (a - 4)(a + 4) = 0$

$(a - 4) = 0 \quad or \quad (a + 4) = 0$

$a - 4 = 0 \quad or \quad a + 4 = 0$

$a = 0 + 4 \quad or \quad a = 0 - 4$

$a = 4 \quad or \quad a = -4$

$Solution\ is\ a = 4\ or\ a = -4$

Solving Quadratic Equations Using the Quadratic Formula:

The Quadratic Formula sates:

$$x = \frac{-b \pm \sqrt{b^2 - 4ac}}{2a}$$

$$where\ a = the\ coefficient\ of\ x^2$$
$$b = the\ coefficient\ of\ x$$
$$and\ c = the\ constanct$$

Again, to use this formula the quadratic equation MUST equal zero first.

Note: This equation can be used to solve the quadratic equation when:

1. Student have difficulty using the factorizing method
2. Whenever the equation CANNOT be factorized.

Example:

1. Solve $2x^2 - x - 15 = 0$

Working:

The above equation can be factorized and solved as seen in Example 2; page 49. But we are going *to use the Quadratic formula to solve instead.*

$$x = \frac{-b \pm \sqrt{b^2 - 4ac}}{2a}$$

From the equation above: $a = 2; b = -1 \ \& \ c = -15$

Hence, substituting for a, b and c into the formula gives:

$$x = \frac{-(-1) \pm \sqrt{(-1)^2 - 4(2)(-15)}}{2(2)}$$

$$x = \frac{1 \pm \sqrt{121}}{4}$$

$$Now: x = \frac{1 + \sqrt{121}}{4} \quad or \quad x = \frac{1 - \sqrt{121}}{4}$$

$$\Rightarrow x = \frac{1 + 11}{4} \quad or \quad x = \frac{1 - 11}{4}$$

$$\Rightarrow x = \frac{12}{4} \quad or \quad x = \frac{-10}{4}$$

$$Hence, \quad x = 3 \quad or \quad x = -2.5$$

$$The\ solution\ is\ x = 3 \quad or \quad x = -2.5$$

2. Solve $5x^2 + 7x - 3 = 0$

Working:

The above equation cannot be factorized and solved as seen. Hence, we will use:

The Quadratic formula to solve instead.

$$x = \frac{-b \pm \sqrt{b^2 - 4ac}}{2a}$$

From the equation above: $a = 5; b = 7 \ \& \ c = -3$

Substituting for a, b and c into the formula gives:

$$x = \frac{-(7) \pm \sqrt{(7)^2 - 4(5)(-3)}}{2(5)}$$

$$x = \frac{-7 \pm \sqrt{109}}{10}$$

$$Now: x = \frac{-7 + \sqrt{109}}{10} \quad or \quad x = \frac{-7 - \sqrt{109}}{10}$$

$$\Rightarrow x = \frac{-7 + 10.44}{10} \quad or \quad x = \frac{-7 - 10.44}{10}$$

$$\Rightarrow x = \frac{3.44}{10} \quad or \quad x = \frac{-17.44}{10}$$

$$Hence, \quad x = 0.344 \quad or \quad x = -1.744$$

$$The\ solution\ is\ x = 0.344 \quad or \quad x = -1.744$$

Activity 20a

Solve the following quadratic equations using the method of factorization

1. $x^2 + 2x = 0$
2. $x^2 + 7x = 0$
3. $x^2 - 9x = 0$
4. $x^2 + 10x + 9 = 0$
5. $x^2 - 5x - 24 = 0$
6. $14 - 5x - x^2 = 0$
7. $x^2 - 3x - 8 = 2$
8. $x^2 + 3x = 10$
9. $-x^2 + 27 = 6x$
10. $x^2 + 2x + 5 = 8$
11. $2x^2 + 5x + 3 = 0$
12. $5x^2 + 16x + 8 = 5$
13. $4x^2 + 11x + 5 = -1$
14. $x^2 - 121 = 0$
15. $3x^2 - 27 = 0$

16. $2x^2 - 50 = 0$ 17. $3x^2 - 108 = 0$ 18. $5x^2 = 125$
19. $16y^2 = 9$ 20. $25a^2 = 4$

Activity 20b

Solve the following quadratic equations using the quadratic formula

1. $2x^2 - 5x + 3 = 0$ 2. $2x^2 + 5x - 9 = 0$ 3. $4x^2 - 7x + 3 = 0$
4. $2x^2 + 13x = 5$ 5. $3x^2 + 5x = 1$ 6. $2x^2 - 6x = -1$

Completing The Square

This method is where the quadratic expression $ax^2 + bc + c$ *is converted to the vertex form* $a(x+h)^2 + k$; *where* $h = \frac{b}{2a}$ *and* $k = \frac{4ac-b^2}{4a}$.

To solve any quadratic equation using this method, we firstly convert the expression in the form $a(x+h)^2 + k$; *then equate to zero then solve.*

Example:

(a) Write $7x^2 + 4x + 8$ *in the form* $a(x+h)^2 + k$

Working:

Note: Since $h = \frac{b}{2a}$ *and* $k = \frac{4ac-b^2}{4a}$ *and from the expression* $7x^2 + 4x + 8; a = 7$, $b = 4$ *and* $c = 8$. *Thefore, substituting into* $h = \frac{b}{2a}$ *and* $k = \frac{4ac-b^2}{4a}$ *to find* h & k

$h = \frac{4}{2(7)} = \frac{2}{7}$ & $k = \frac{4(7)(8)-(4)^2}{4(7)} = \frac{52}{7} = 7\frac{3}{7}$

Now, substituting for h, k *and* a *in the form* $a(x+h)^2 + k$ *gives*: $7\left(x + \frac{2}{7}\right)^2 + 7\frac{3}{7}$

Solving Quadratic Equations - Completing The Square Method

Note: If asked to solve a quadratic equation using completing the square method. We do the following:

- First, make the equation equal to zero.
- We then write the expression in the form $a(x+h)^2+k$ and equate to zero.
- Finally, we use the concept of transposition and solving equation to solve.

Example:

Solve the quadratic equation below by using the completing the square method

$$5x^2+10x-2=0$$

Working: $Completing\ the\ square\ format\ is\ a(x+h)^2+k; where\ h=\frac{b}{2a}\ and$

$$k=\frac{4ac-b^2}{4a}$$

$Given\ that\ from\ the\ equation\ a=5;\ \ b=10\ \ \ and\ c=-2; hence\ substituting$

$$h=\frac{b}{2a}=\frac{(10)}{2(5)}=1 \quad and \quad k=\frac{4(5)(-2)-(10)^2}{4(5)}=-7$$

Hence: $substituting\ for\ a, h\ and\ k\ into\ the\ format\ a(x+h)^2+k\ gives$

$\Rightarrow 5(x+1)^2+(-7)=0$

$\Rightarrow 5(x+1)^2-7=0$

Now, solving for the values of x

$\Rightarrow 5(x+1)^2=0+7$

$\Rightarrow 5(x+1)^2=7$

$\Rightarrow (x+1)^2=\frac{7}{5}$

$\Rightarrow (x+1)^2=1.4$

$\Rightarrow x+1=\pm\sqrt{1.4}$

$\Rightarrow x+1=\pm 1.18$

$\Rightarrow x=-1\pm 1.18$

$\Rightarrow x=-1+1.18 \quad or \quad x=-1-1.18$

Solutions are $x=0.18 \quad or \quad x=-2.18$

Activity 20c

Write each of the following in the $form\ a(x+h)^2 + k\ and\ solve\ 6\ and\ 7\ using\ completing\ the\ square\ method.$

1. $4x^2 + 5x + 7$ 2. $-2x^2 + 5x + 1$ 3. $6x^2 - 8x + 7$ 4. $5 - 12x + 3x^2$
5. $-7 - 6x + 8x^2$ 6. $4x^2 - 7x - 5 = 0$ 7. $-3x^2 - 10x = -8$

Transposition 2 – Subject appear more than once

Note: Before we can transpose for a subject that appear more than once, we MUST factorize. That is, we gather the subject together and factor out that subject which makes it appearance to be once; then we can transpose.

Example:

1. Given $xp + yp = c; make\ p\ the\ subject$

Working: $p\ appear\ twice\ so\ we\ have\ to\ factorize\ by\ factoring\ out\ the\ p.$

$p(x+y) = c$

$\Rightarrow p = \frac{c}{(x+y)}$

2. $If\ bx = cx + d;\ for\ x$

Working: $x\ appear\ twice\ so\ we\ have\ to\ factorize$ But we have to get the terms with the x on the same side

$bx - cx = d$

$x(b-c) = d$

$\Rightarrow x = \frac{d}{(b-c)}$

Activity 21

1. $Given\ ab - bc = xy;\ make\ b\ the\ subject\ of\ the\ formula.$
2. $If\ P + Prt = x;\ make\ P\ the\ subject\ of\ the\ formula.$
3. $Given\ 3x - xy + n;\ make\ x\ the\ subject\ of\ the\ formula.$
4. $Given\ V(R - M) = R;\ make\ R\ the\ subject\ of\ the\ formula.$
5. $If\ \frac{mx-y}{n} = x;\ make\ x\ the\ subject\ of\ the\ formula.$

Compound Inequality

Example 1: *Represent the following inequalities on a number line*

1. $-2 < x < 3$

Working:

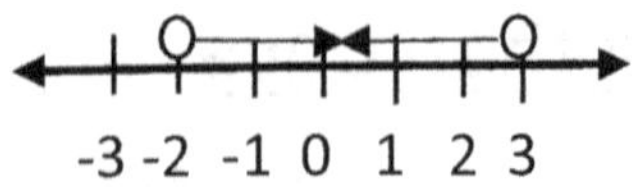

2. $-1 \leq x \leq 2$

Working:

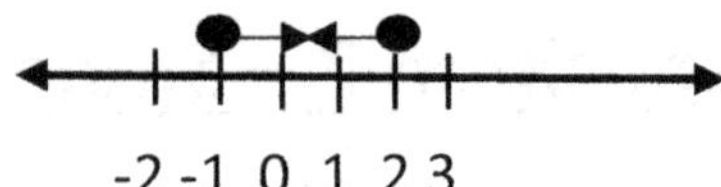

Example 2:

(a) *Given that x is an integer, find the solution set of x for the inequality* $5 \leq 2x + 1 < 11$

(b) *Represent your solution on a number line.*

Working:

$5 \leq 2x + 1 < 11$ *Removing the + 1 to both side of the two inequality signs gives:*

$5 - 1 \leq 2x < 11 - 1$

$4 \leq 2x < 10$ *Now, removing the coefficient 2 from the 2x to both sides of the two inequality signs gives:*

$\frac{4}{2} \leq x < \frac{10}{2}$

$2 \leq x < 5$

(b) 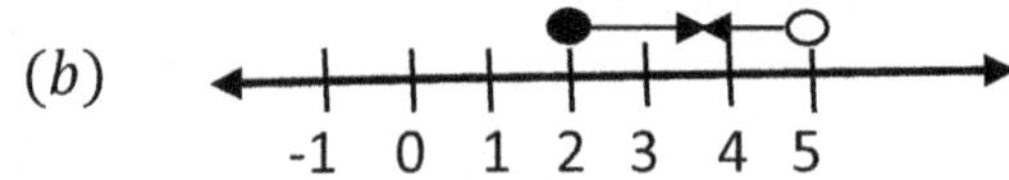

Example 3:

Given that x is an integer, find the solution set of x for the inequality

$$-1 \leq 7 - 4x \leq 11$$

Working:

$-1 \leq 7 - 4x \leq 11$ *Removing the + 7 to both side of the two inequality signs gives:*

$-7 - 1 \leq -4x \leq 11 - 7$

$-8 \leq -4x \leq 4$ *Now, removing the coefficient − 4 from the − 4x to both sides of the two inequality signs.*

$\frac{-8}{-4} \geq x \geq \frac{4}{-4}$ *However*, the direction of the inequality sign changes because we are dividing by a negative coefficient.

$2 \geq x \geq -1$

Activity 22a

Represent the following inequalities on a number line

1. $-1 < x \leq 3$ 2. $-1 \leq x \leq 2$ 3. $0 < x < 4$
4. $-5 \leq x \leq -1$ 5. $1 \leq x \leq 4$

Activity 22b

Solve the following inequalities

1. $2 < 2x < 6$ 2. $-1 \leq 9 + x \leq 17$ 3. $-50 < 7x + 6 < -8$
4. $-3 \leq \frac{x}{2} < 0$ 5. $2 < -2x < 6$ 6. $4 < 4 - 2x < 8$

Non-Linear Simultaneous Equations

These equations are where at least one of the given equations is non-linear.

Note: All linear equations produce a straight line on a graph.

Example:

Solve the following simultaneous equations

$x - y = 6$ ⟶ (Linear equation)

$xy = 16$ ⟶ (Non-linear equation)

Working:

To solve we have to use the substitution method, because we have a non − linear equation $(xy = 16)$.

Hence:

Using equation 1, and making *x the subject gives:* $x = 6 + y$.

Now, substituting for $x = 6 + y$ *into equation 2;* $xy = 16$ *for x gives:*

$(6 + y)y = 16$ *rewritten as* $y(6 + y) = 16$

Removing the braket gives $6y + y^2 = 16$

Note: *By observation, we can see a quadratic equation has developed.*

Note: *Remember to solve quadratic equation, the equation must equate to zero before solving using any of the methods, stated. We will use the factorization method, since equation can be factorized.*

$\Rightarrow y^2 + 6y - 16 = 0$

When factorize gives:

$(y - 2)(y + 8) = 0$

Now: $y - 2 = 0$ *and* $y + 8 = 0$

$y = 0 + 2$ *and* $y = 0 - 8$

$$y = 2 \qquad and \qquad y = -8$$

$Now, solving\ for\ x\ when\ y = 2\ and\ y = -8\ using\ equation\ 1;\ x - y = 6$

When:

$y = 2\ then\ x - (2) = 6 \qquad and \qquad y = -8\ then\ x - (-8) = 6$

$x - 2 = 6 \qquad and \qquad x + 8 = 6$

$x = 6 + 2 \qquad and \qquad x = 6 - 8$

$x = 8 \qquad and \qquad x = -2$

Therefore, the solutions are:

$$x = 8\,; y = 2 \qquad and \qquad x = -2; y = -8$$

Activity 23

Solve each of the pairs of simultaneous equations:

1. $xy = 24$
   $x - y = 10$
2. $x - y = -5$
   $xy = 6$
3. $x + y = 11$
   $xy = 24$
4. $x + y = 4$
   $x^2 + y = 6$
5. $y + 3x = 13$
   $xy = 4$

QUIZ 1

Answer All the questions

1. $Given\ a = 4, b = 2, c = -3\ and\ d = 0, evaluate$

   $(i)\ 5a - 3c \quad (ii) abc \quad (iii)\ \frac{1}{2}ab - 12d$

2. $Simplify$

   $(a)\ 3xy - 7 - 2xy + 15$ $\qquad$ $(b)\ 15m - 2(4 + 3m) + 10$

3. Translate the following word statements into algebraic expressions

   $(a)\ four\ times\ a\ number\ y, plus\ seventeen$

   $(b)\ the\ square\ of\ the\ sum\ of\ two\ numbers, x\ and\ five$

4. Solve

   $(a)\ 9 = x + 3$ $\qquad$ $(b)\ 2(x - 5) = 17$ $\qquad$ $(c)\ \frac{x}{3} - \frac{4}{7} = \frac{x}{7}$

5. Simplify

   $(a)\ \frac{x}{9} + \frac{x}{3}$ $\qquad$ $(b)\ 2a^5 \times 3 \times 4a^2$ $\qquad$ $(c)\ 16y^9 \div 8y$

6. Evaluate

   $(3)^2$ $\qquad$ $(b)\ 7^0$ $\qquad$ $(c)\ 2^{-3}$

7. Factorize completely

   $(a)\ 7y + 14$ $\qquad$ $(b)\ 12a^2b - 10ab^2 + 16a^3b^2$

8. $Given\ A = \frac{3x+y}{m}$

   $(a)\ find\ the\ value\ of\ A\ when\ x = 2, y = 8\ qnd\ m = 10$

   $(b)\ transpose\ the\ formula\ to\ make\ y\ the\ subject$

9. Solve the following pair of simultaneous equations

   $7x - 2y = 19$

   $3x = 14 - 5y$

QUIZ 2

Answer All questions

1. Simplify

   $(a)\ (3a^2)^3$ $(b)\ (16x^{10})^{\frac{3}{2}}$

2. Simplify

   $(a)\ \frac{x-4}{4} - \frac{x+3}{8}$ $(b)\ \frac{2}{x} + \frac{7}{x-3}$ $(c)\ \frac{1}{x-1} + \frac{2}{(x+5)(x-1)}$

   $(d)\ \frac{4}{5x} - \frac{2}{3x^2}$

3. Solve for x

   $4^{2x} = 8$

4. Factorize

   $(a)\ ab + 3a + 2b + 6$ $(b)\ 2x^2 + 13x - 7$ $(c)\ x^2 - 9$

5. Write as a single fraction in simplest form

   $(i)\ \frac{a^2b}{x^2} \times \frac{3x^5y}{a^4b^2}$ $(b)\ \frac{18y}{7x^2} \div \frac{3xy}{2q}$

6. Solve

   $(a)\ 3 \geq 3 - 5x > -12$ $(b)\ \frac{3x}{4} + \frac{21}{4} \geq \frac{2x}{3} + 5$

7. $Make\ x\ the\ subject\ of\ the\ formula\ where\ xy = xp + mn$

8. Solve

   $x^2 + 5x - 14 = 0$

9. $Write\ x^2 + 9x - 5\ in\ the\ format\ a(x+h)^2 + k$

10. $Solve\ the\ following\ pairs\ of\ simultaneous\ equations$

    $(a)\ xy = 40$ $(b)\ x^2 + y = 6$

    $x - y = -3$ $x - y = -4$

11. *Simpligy*

$(a)\ 7a^2(a^2 - a^{-1})$ $(b)\ (5n^{-4}y^3)(2n^7y^{-1})$

12. *Solve*

$$\frac{5x+2}{12} = \frac{4x-3}{5}$$

13. *Solve*

$$\frac{1}{x-3} - \frac{x-5}{3} = 0$$

Key:

Activity 1a

1. 5y
2. $\frac{cd}{4}$ or $\frac{1}{4}cd$
3. $(x + y)^2$
4. $10x + y$
5. $8p - q$
6. $(x - y)^3$
7. $\frac{2x+6y}{5z}$
8. $(ab)^2 - \sqrt{c}$
9. $\sqrt{(x + y)}$

Activity 1b

1. Seven times a number y
2. four times a number m, plus 5
3. Nine times a number y, subtract 8
4. The square of the sum of x and 3y
5. Two thirds the product of **a** and **b**
6. The sum of two numbers **a** and **b** divide by 5

Activity 1c

1. (a) $52 + y$ (b) $y - 17$
2. (a) $n + 6$ (b) $3(n - 7)$
3. (i) $4f$ (ii) $f - 7$
4. (a) $95g$ (b) $g + 34$
5. (a) $\frac{1}{3}b$ (b) $b - 12$
6. (a) $k - 7$ (b) $\frac{15+k}{4}$

Activity 2a

1. 5
2. 119
3. 0
4. 49
5. 13
6. 1
7. (a) 2.5 (b) 148
8. 85

Activity 2b

1. 1050
2. 25
3. 2.9 to I decimal place
4. 71.5
5. 21
6. 52.1
7. 2000
8. 34.34

Activity 3

1. $10n$
2. $12x$
3. $5x$
4. $-5x$
5. $-2p$
6. $9r$
7. $7x$
8. $12n^2$
9. $20x - 2p$
10. $-2r^2 + 2r$
11. $3x^2 - 4y^2$
12. $2p - 9$
13. $13a^2b - 6ab^2 - 5$
14. $\frac{3}{8}y - \frac{1}{4}x$
15. $5xy - 10x^2y - 1.1xy^2$
16. $4gh - 1h$
17. $5ab + 1a$

Activity 4

1. a^6
2. $6m^8$
3. $16x^4y^6$
4. $0.63a^3b^4c^2$
5. $\frac{1}{2}m^3n^4$
6. $8x^5y^3$
7. $42p^3q^2$
8. $3y^4$
9. $0.2x$ or $\frac{1}{5}x$
10. $8x^2y^6$
11. $\frac{1.5b^2}{c}$
12. $15m^4n^2$
13. $\frac{-25q^2}{p}$
14. $2a^3$
15. $\frac{2}{3}m^2n$
16. $3r^2yz$

Activivirty 5a

1. $21x + 7y$
2. $-4a - 8b$
3. $-2n + 4m$
4. $2 - 6x$
5. $-a - 2b$
6. $2a + 1$
7. $1a - 4b$
8. $1.5 - 6x$

Activity 5b

1. $6a + 10b$
2. $2x - 8y$
3. $7n + 13m$
4. $-1x + 16y$
5. $1x + 5$
6. $8 + 4y$
7. $1 + 6y$
8. $-13 + 1x$
9. $23 + 1x$

Activity 6a

1. $2(a + 2)$
2. $x(m - n)$
3. $6(3a - c)$
4. $4(3x - 4)$
5. $3ab(1 + 2c)$

Activity 6b

1. $9xy(3x - 1)$
2. $3ab(2a^2 + b)$
3. $15a^2(a - 2)$
4. $7mn(m - 3)$
5. $6ab(3a - 2b)$
6. $6b(2ab + 3a^2 - 1)$
7. $ab(3a - 6 + 5b)$
8. $mn(4n + 1 + 3m)$
9. $7y(5y^2 - 1 + 2y)$
10. $4y(3x^2 + 2xy - 1)$

Activity 7

1. $\frac{7y}{20}$ 2. $\frac{23a}{30}$ 3. $\frac{x}{3}$ 4. $\frac{7xy}{10}$
5. $\frac{1y}{8}$ 6. $\frac{11m}{15}$ 7. $1\frac{3x}{10}$ 8. $-\frac{14y}{15}$

Activity 8a

1. $x = 8$ 2. $x = 5$ 3. $x = -6$ 4. $x = 2.4$
5. $x = 40$ 6. $x = 4$ 7. $x = 15$ 8. $x = 36$
9. $x = 14$ 10. $x = 2$ 11. $x = 5$ 12. $x = 5$
13. $x = 1$ 14. $x = -6.5$ 15. $x = 28$ 16. $x = 8$

Activity 8b

1. $x = 5$. 2. $x = 3$ 3. $x = 4$ 4. $x = 4$ 5. $x = 7$
6. $x = 2$ 7. $x = 7$ 8. $y = 4$ 9. $x = 2.8$ 10. $x = 6$

Activity 9

1. $y = -3$ 2. $y = 30$ 3. $x = -5$ 4. $x = 2.25$ 5. $x = \frac{5}{6}$ 6. $c = -1.5$
7. $x = 2\frac{1}{4}$

Activity 10a

1. $x < 19$ 2. $y > -1$ 3. $p \leq 25$ 4. $q \geq 12$

Activity 10b

1. $x < 2$ 2. $x \geq -5$ 3. $x \leq 9$ 4. $x > -1$ 5. $x \leq -3$ 6. $x \leq -5$
7. $x < 1$ 8. $x > 1$ 9. $x \leq 6$ 10. $x > 15$ 11. $x \leq -1.5$ 12. $x \leq 1$

Activity 11

1. $x = 2$ & $y = -3$ 2. $x = 7$ & $y = -2$ 3. $x = 5$ & $y = 4$
4. $x = -1.25$ & $y = -2$ 5. $x = 3$ & $y = -2$ 6. $x = 1.5$ & $y = 7$
7. $x = 10$ & $y = 15$ 8. $x = 3$ & $y = -1$ 9. $x = 5$ & $y = 4$
10. $(a)\ x + y = 17$ $(b)(i) 0.10x$ $(ii) 0.25y$ $(c)\ 0.10x + 0.25y = \$3.05$
$(d)\ x = 8$ & $y = 9$

Activity 12a

1. 2^4 2. 4^6 3. y^3 4. a^5 5. 2^{10} 6. 3^{11}
7. x^4y^6 8. 6^5 9. 4^7 10. p^4 11. a^5b^2 12. xy^6

Activity 12b

1. 64 2. 729 3. $4a^6$ 4. $27n^6b^3$ 5. $\frac{256}{729}$ 6. $\frac{1}{625}$ 7. $\frac{y^8}{x^{10}}$ 8. $\frac{256}{y^{10}}$
9. $\frac{b^8}{x^2y^{10}}$ 10. $\frac{x^4y^6}{n^{10}}$

Activity 12c

1. 1 2. 1 3. 1 4. 1 5. 1

Activity 12d

1. $\frac{1}{5^3}$ 2. $\frac{1}{3^2}$ 3. $\frac{1}{7^2}$ 4. $\frac{1}{x^2}$ 5. $\frac{1}{y^4}$ 6. $\frac{b^2}{a^3}$

Activity 12e

1. 2 2. 4 3. 9 4. 12 5. 1 6. 1 7. 3 8. 2 9. 5

10. 16 11. 16 12. $25m^2$ 13. $216y^6$ 14. $9y^6$ 15. $64n^9$

Activity 13

1. $m = \frac{y-c}{x}$ 2. $t = \frac{mx}{4h^2}$ 3. $h = \frac{A-3x^2}{4y}$ 4. $y = \frac{2x}{T}$
5. $a = \frac{v^2-u^2}{2x}$ 6. $c = \sqrt{\frac{ab}{d}}$ 7. $b = \frac{c}{x} - a$ or $b = \frac{c-ax}{x}$

8. $t = \frac{A-C}{Cr}$ 9. $b = \frac{2A}{h} - a$ or $b = \frac{2A-ah}{h}$ 10. $u = \sqrt{v^2 - 2ax}$

Activity 14

1. $(x+y)(m+n)$ 2. $(a+c)(b+d)$ 3. $(x+y)(a+5b)$ 4. $(x-4y)(2a+b)$
5. $(a-4n)(p+q)$ 6. $(r+t)(4x-5y)$ 7. $(x-y)(2a-3b)$ 8. $(x-y)(3r-5b)$

Activity 15

1. x^2+5x+6 2. x^2+7x+6 3. x^2-6x-7 4. x^2+6-16
5. $x^2-7x+10$ 6. $x^2-7x+12$ 7. $2x^2+7x+3$
8. $2x^2+2x-12$ 9. $x^2+10x+25$ 10. x^2-6x+9
11. $4x^2+4x+1$ 12. $4-8x+4x^2$

Activity 16a

1. $3x(4+3x)$ 2. $2x(3x-4)$ 3. $6x(5x+3)$ 4. $4x(2x+5)$
5. $5x(3-4x)$ 6. $7x(1+2x)$

Activity 16b

1. $(x+4)(x+3)$ 2. $(x+4)(x+5)$ 3. $(x-7)(x-6)$ 4. $(x-6)(x-5)$
5. $(x-1)(x+4)$ 6. $(x-9)(x+8)$ 7. $(2+x)(3-x)$ 8. $(3-4x)(1+2x)$
9. $(4x+3)(3x+1)$ 10. $(5-7x)(2-3x)$

Activity 16c

1. $(3x+4)(5x+2)$ 2. $(4x+1)(4x+1)$ *or* $(4x+1)^2$ 3. $(3x+2)(4x-3)$
4. $(x-2)(2x-3)$ 5. $(3x+1)(x-4)$ 6. $(3x-1)(2x+3)$
7. $(3x+1)(x-2)$ 8. $(2x+1)(4x+3)$ 9. $(3x+1)(5x-3)$
10. $(3x-4)(2x+5)$

Activity 16d

1. $(x-12)(x+12)$ 2. $(x-1)(x+1)$ 3. $(x-10)(x+10)$ 4. $(x-7)(x+7)$
5. $(x-9)(x+9)$ 6. $(3x-1)(3x+1)$ 7. $(5x-1)(5x+1)$
8. $(6x-1)(6x+1)$ 9. $(2x-3)(2x+3)$ 10. $(3x-6)(3x+6)=9(x-2)(x+2)$

Activity 17

1. $\frac{41}{35x}$ 2. $\frac{17}{45x}$ 3. $\frac{14+15x}{24x}$ 4. $\frac{15+4x}{10x}$ 5. $\frac{27-8y}{15xy}$

6. $\frac{10x-12}{15x^2}$ 7. $\frac{12+5y}{14y^2}$ 8. $\frac{8m-15}{6mn}$ 9. $\frac{7x+17}{12}$ 10. $\frac{x+6}{12}$

11. $\frac{x+1}{12}$ 12. $\frac{9-x}{8}$ 13. $\frac{10x+35}{(x)(x+5)}$ 14. $\frac{3x+6}{(x)(x+3)}$ 15. $\frac{13x-2}{(x+2)(x-5)}$

16. $\frac{4x+8}{(x)(x-2)}$ 17. $\frac{4x-13}{(x-5)(x-4)}$ 18. $\frac{-1x-21}{(x+1)(x-3)}$ 19. $\frac{-5x+23}{(x-1)(x+5)}$

20. $\frac{5x+60}{(x)(x+5)}$ 21. $\frac{7x+23}{(x+1)(x+2)}$ 22. $\frac{-3x+25}{(x-7)(x-2)}$ 23. $\frac{12x+31}{(x+3)(x-6)}$

24. $\frac{1x+11}{(x-7)(x+7)}$ *or* $\frac{1x+11}{x^2-49}$

Activity 18a

1. $\frac{1mx}{4}$ 2. $\frac{x^2b}{y}$ 3. $\frac{c}{a^2b}$ 4. . $\frac{p^3}{qr}$ **5.** $\frac{ax^2}{bc}$

6. mnq 7. $\frac{2am}{bn^2}$ 8. $\frac{10m^2}{nr}$

Activity 18b

1. $\frac{(a-2)}{5}$ 2. $\frac{(x+3)}{5}$ 3. $\frac{7}{(x-4)}$ 4. $\frac{3}{(y-3)}$ 5. $\frac{3}{(x+2)}$ 6. $\frac{3}{(x+5)}$

Activity 19

1. $x = 6$ 2. $x = 0$ 3. $x = 1.5$ 4. $x = 2.5$ 5. $x = -1.5$ 6. $x = -7$
7. $x = -1$ 8. $x = 4$

Activity 20a

1. $x = 0$ or $x = -2$ 2. $x = 0$ or $x = -7$ 3. $x = 0$ or $x = 9$
4. $x = -1$ or $x = -9$ 5. $x = 8$ or $x = -3$ 6. $x = 2$ or $x = -7$
7. $x = 5$ or $x = -2$ 8. $x = 2$ or $x = -5$ 9. $x = 3$ or $x = -9$
10. $x = 1$ or $x = -3$ 11. $x = -\frac{3}{2}$ or $x = -1$ 12. $x = -3$ or $x = -\frac{1}{5}$
13. $x = -\frac{3}{4}$ or $x = -2$ 14. $x = 11$ or $x = -11$ 15. $x = 3$ or $x = -3$
16. $x = 5$ or $x = -5$ 17. $x = 6$ or $x = -6$ 18. $x = 5$ or $x = -5$
19. $y = \frac{3}{4}$ or $y = -\frac{3}{4}$ 20. $a = \frac{2}{5}$ or $a = -\frac{2}{5}$

Activity 20b

1. $x = 1$ or $x = 1.5$ 2. $x = -3.71$ or $x = 1.21$ 3. $x = 0.75$ or $x = 1$
4. $x = 0.36$ or $x = -6.86$ 5. $x = 0.18$ or $x = -1.85$ 6. $x = 0.18$ or $x = 2.82$

Activity 20c

1. $4\left(x+\frac{5}{8}\right)^2 + 5\frac{7}{16}$ 2. $-2\left(x-\frac{5}{4}\right)^2 + 4\frac{1}{8}$ 3. $6\left(x-\frac{2}{3}\right)^2 + 4\frac{1}{3}$

4. $3(x-2)^2 - 7$ 5. $8\left(x-\frac{3}{8}\right)^2 - 8\frac{1}{8}$

6. $x = 2.295$ or $x = -0.545$ 7. $x = \frac{2}{3}$ or $x = -4$

Activity 21

1. $b = \frac{xy}{(a-c)}$ 2. $P = \frac{x}{(1+rt)}$ 3. $x = \frac{n}{(3-y)}$ 4. $R = \frac{VM}{(V-1)}$

5. $x = \frac{y}{(m-n)}$

Activity 22a

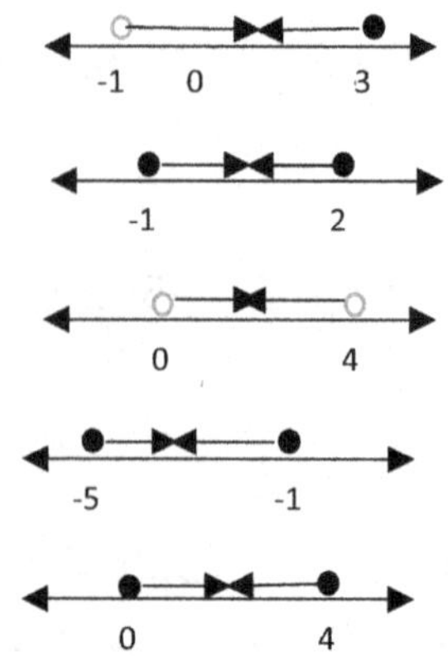

Activity 22b

1. $1 < x < 3$ 2. $-10 \leq x \leq 8$ 3. $-8 < x < -2$ 4. $-6 \leq x < 0$

5. $-1 > x > -3$ 6. $0 > x > -2$

Activity 23

1. $(12,\ 2)$ *and* $(-2, -12)$ 2. $(-6, -1)$ *and* $(1,\ 6)$ 3. $(3,\ 8)$ *and* $(8,\ 3)$
4. $(-1,\ 5)$ *and* $(2,\ 2)$ 5. $\left(\frac{1}{3},\ 12\right)$ *and* $(4,\ 1)$

Quiz 1

1. $(i)\ 29\ \ (ii) - 24\ \ (c)\ 4$
2. $(a)\ xy + 8\ \ (b)\ 9m + 2$
3. $(a)\ 4y + 17\ \ (b)\ (x + 5)^2$
4. $(a)\ x = 6\ \ (b) x = 13.5\ \ (c)\ x = 3$
5. $(a)\ \frac{4x}{9}\ \ (b)\ 24a^7\ \ (c)\ 2y^8$
6. $(a)\ 9\ \ (b)\ 1\ \ (c)\ \frac{1}{8}$
7. $(a)\ 7(y + 2)\ \ (b)\ 2ab(6a - 5b + 8a^2b)$
8. $(a)\ A = 1.4\ \ \ (b)\ y = Am - 3x$
9. $x = 3$ *and* $y = 1$

Quiz 2

1. $(a)\ 27a^6 \quad (b)\ 64x^{15}$
2. $(a)\ \dfrac{x-11}{8} \quad (b)\ \dfrac{9x-6}{(x)(x-3)} \quad (c)\ \dfrac{x+7}{(x+5)(x-1)} \quad (d)\ \dfrac{12x-10}{15x^2}$
3. $x = \dfrac{3}{4}$
4. $(a)\ (b+3)(a+2) \quad (b)\ (2x-1)(x+7) \quad (c)\ (x-3)(x+3)$
5. $(i)\ \dfrac{3x^3y}{a^2b} \quad (ii)\ \dfrac{12q}{7x^3}$
6. $(a)\ 0 \le x < 3 \qquad (b)\ x \ge -3$
7. $x = \dfrac{mn}{(y-p)}$
8. $x = 2 \ \ or \ \ x = -7$
9. $2\left(x+\dfrac{9}{2}\right)^2 - \dfrac{101}{4}$
10. $(a)\ (5,\ 8)\ and\ (-8,-5) \qquad (b)\ (1,\ 5)\ and\ (-2,\ 2)$
11. $(a)\ 7a^4 - 7a \qquad (b)\ 10n^3y^2$
12. $x = 2$
13. $x = 2 \ \ or \ \ x = 6$

Index

Made in the USA
Columbia, SC
28 February 2025